CAMBRIDGE LIBRARY COLLECTION

Books of enduring scholarly value

Earth Sciences

In the nineteenth century, geology emerged as a distinct academic discipline. It pointed the way towards the theory of evolution, as scientists including Gideon Mantell, Adam Sedgwick, Charles Lyell and Roderick Murchison began to use the evidence of minerals, rock formations and fossils to demonstrate that the earth was older by millions of years than the conventional, Bible-based wisdom had supposed. They argued convincingly that the climate, flora and fauna of the distant past could be deduced from geological evidence. Volcanic activity, the formation of mountains, and the action of glaciers and rivers, tides and ocean currents also became better understood. This series includes landmark publications by pioneers of the modern earth sciences, who advanced the scientific understanding of our planet and the processes by which it is constantly re-shaped.

Researches in Theoretical Geology

The geologist Sir Henry Thomas De la Beche (1796–1855) made important contributions as both a surveyor and a theorist. Elected to the Royal Society in 1823, he mapped geological strata in Devon during the 1830s and became the founding director of the British Geological Survey, the world's first national geological survey. In 1847, he was elected president of the Geological Society of London. Reflecting the scope of his scientific knowledge, the present work covers a wide range of topics, including the density of planets, the mineralisation of organic remains, and what could be inferred from the fossils thus created. The book was first published in 1834, the year he became embroiled in an argument with his contemporary Roderick Murchison. Lasting several years, the dispute became known as the 'The Great Devonian Controversy'. De la Beche's *Geological Manual* (third edition, 1833) has also been reissued in this series.

Cambridge University Press has long been a pioneer in the reissuing of out-of-print titles from its own backlist, producing digital reprints of books that are still sought after by scholars and students but could not be reprinted economically using traditional technology. The Cambridge Library Collection extends this activity to a wider range of books which are still of importance to researchers and professionals, either for the source material they contain, or as landmarks in the history of their academic discipline.

Drawing from the world-renowned collections in the Cambridge University Library and other partner libraries, and guided by the advice of experts in each subject area, Cambridge University Press is using state-of-the-art scanning machines in its own Printing House to capture the content of each book selected for inclusion. The files are processed to give a consistently clear, crisp image, and the books finished to the high quality standard for which the Press is recognised around the world. The latest print-on-demand technology ensures that the books will remain available indefinitely, and that orders for single or multiple copies can quickly be supplied.

The Cambridge Library Collection brings back to life books of enduring scholarly value (including out-of-copyright works originally issued by other publishers) across a wide range of disciplines in the humanities and social sciences and in science and technology.

Researches in Theoretical Geology

Henry T. De La Beche

CAMBRIDGE
UNIVERSITY PRESS

University Printing House, Cambridge, CB2 8BS, United Kingdom

Published in the United States of America by Cambridge University Press, New York

Cambridge University Press is part of the University of Cambridge.
It furthers the University's mission by disseminating knowledge in the pursuit of education, learning and research at the highest international levels of excellence.

www.cambridge.org
Information on this title: www.cambridge.org/9781108066969

This edition first published 1834
This digitally printed version 2014

ISBN 978-1-108-06696-9 Paperback

THE EARTH

H.T. de la Beche del.   H. Adlard sculp.

Supposed to be seen from Space

RESEARCHES

IN

THEORETICAL GEOLOGY.

BY

H. T. DE LA BECHE, F.R.S., V.P.G.S.

MEMB. GEOL. SOC. OF FRANCE, CORR. MEMB. ACAD. NAT. SCI. PHILADELPHIA, ETC.

LONDON:

CHARLES KNIGHT, 22, LUDGATE STREET.

1834.

Printed by Richard Taylor, Red Lion Court, Fleet Street.

PREFACE.

ALTHOUGH the theory of central heat and the former igneous fluidity of our planet have been much dwelt upon in the following pages, the author trusts that he will not be considered so attached to these views as not to be ready to reject them and embrace others which may afford a better explanation of an equal number of observed facts, should such be brought forward. It can only be amid a thousand errors, and by a determination to abandon our preconceived opinions, when shown to be untenable, not by pertinaciously adhering to them because we have once adopted them, that we can approximate towards the truth. By strictly advocating a particular theory, prominently displaying those facts only which may appear to afford it support, we are in perpetual danger of deceiving ourselves and others. Facts of all kinds, whether in favour of or against our views, should be honestly brought forward, in order that those whose opinions are unprejudiced may fairly weigh the evidence adduced.

As when we approach simplicity we generally approach truth, the author has in the following researches been always anxious to seek for a simple cause, explanatory of several facts, such as the former igneous fluidity of the earth, rather than adopt other views which may have more novelty though not the same simplicity to recommend them. In doing so he may, indeed, not always have succeeded in referring effects to the right causes, but he trusts, should his conclusions be found erroneous, that he will not be accused of any unfair advocacy of opinions, suppressing or only half stating facts for the purpose of rendering any given hypothesis apparently more tenable than another.

CONTENTS.

CHAPTER I.

CHAPTER II.

CHAPTER III.

CHAPTER IV.

CHAPTER V.

CHAPTER VI.

CHAPTER VII.

CHAPTER VIII.

CHAPTER IX.

CHAPTER X.

CHAPTER XI.

CHAPTER XII.

CHAPTER XIII.

CHAPTER XIV.

CHAPTER XV.

CHAPTER XVI.

CHAPTER XVII.

CHAPTER XVIII.

APPENDIX.

RESEARCHES

IN

THEORETICAL GEOLOGY.

CHAPTER I.

THE sun and the known planets of our solar system are of different densities; it therefore follows that the materials of which these bodies are respectively formed are either different, or do not exist under precisely the same conditions in each. Hence a given density is not necessary to the existence of a planet; and, consequently, there is no argument, *à priori*, against the supposition that the density of a planet, such as the earth, may have changed during the lapse of time.

There is no direct evidence to prove that the matter of the other planets either differs from, or is the same with, that of the earth. It can only be assumed that, in accordance with the simplicity and grandeur of design manifested in the Creation, the general character of such matter would not materially differ, precisely as we see an infinite variety of form produced by the combination of a few elementary substances in animals and vegetables.

Assuming therefore—and it does not appear unphilosophical to do so—that the matter of the planets does not materially differ as respects its general character, there must be some agent or force counteracting the effect of gravity unequally in each of these bodies; otherwise a large planet, such as Saturn, would not be less dense than the earth. When such an agent is sought, heat readily presents itself as capable of producing the effects we endeavour to explain. It is capable of changing the density of all bodies, and we have every reason to believe that all ponderable matter would become gaseous, if the heat applied to it were sufficiently intense. The effect of gravity on the surface of the sun has been estimated to be so great, that a man, if he could be transported there, would be crushed by his own weight. Yet the density of the sun is comparatively inconsiderable. It has therefore been inferred that great heat exists in the interior, enabling it to resist the enormous pressure exerted upon it*. If this mode of reasoning be applicable to the sun, it would also appear applicable to large planets, such as Saturn, the density of which is considered not much to exceed that of cork†. Now, if it be probable that considerable heat is an antagonist force to the pressure, from the action of gravity, of the matter of Saturn towards its centre, preventing that density which it would otherwise possess, there seems no reason, *à priori*, why considerable heat should not exist in the interior of the earth, modifying its density. Hence, if this mode of viewing the subject be correct, and assuming that the

* Herschel, Treatise on Astronomy, p. 239. † Ibid. p. 278.

matter forming our planet is, on the whole, much the same as that composing another, variations in the density of the planets are greatly due to the different intensities of heat in each.

That many of the planets, at least all comparatively near the sun, possess atmospheres, seems nearly certain. Assuming this to be true, we have proof that matter exists at least in two states in these bodies. There is, therefore, no great difficulty in advancing a step further, and in considering that matter may exist in the solid, liquid and gaseous state in all the planets. How far this may also be the case with those minor bodies termed satellites is another question. It is commonly supposed that our satellite, the moon, has no atmosphere; there is, however, some difficulty, when we regard the volcanic character of its solid surface, in considering this absolutely true. Sir John Herschel has observed appearances on the moon's surface which justify him in concluding, that on some of the lunar mountains there are "decisive marks of volcanic stratification, arising from successive deposits of ejected matter*." Now the volcanic eruptions of the moon must be very different from those on the surface of the earth, if they be not accompanied by the evolution of gas. If gases be evolved in lunar volcanic eruptions, gravity necessarily brings them down on the moon's surface, and they can only disappear from thence, either by combining with liquid or solid matter, by the influence of intense cold, or by the effect of considerable pressure. Pressure on the moon's surface can only arise from the attraction

Herschel, Treatise on Astronomy, p. 229.

of the moon itself, and therefore must be altogether insufficient to produce the effect required. Combination with liquid or solid matter must necessarily depend upon chemical affinities, so that, if there be any general resemblance of the matter of the moon to the matter of the earth, we should consider that some of the gases evolved would at least require time to combine with the superficial liquids and solids. If, therefore, volcanic eruptions were moderately frequent, gases would be found on the moon's surface, provided the temperature were not too low, on the same surface, for more than their momentary existence.

If there be any truth in the calculations of Baron Fourier, that the planetary spaces have a temperature $= -58°$ Fahr., comparatively small bodies, such as the moon, would radiate their heat sooner than large bodies, such as Saturn, supposing the various planets and their satellites to have been formed at the same period,—to have possessed an equal temperature, more elevated than that of the surrounding planetary space,—and to be composed of similar matter. Upon this hypothesis the moon and the earth could not long preserve equal temperatures, the former becoming colder than the latter; so that great cold might exist on the moon, while the earth was comparatively warm. We might, viewing the subject in this light, suppose the temperature on the surface of the moon to be so depressed as to prevent the existence of an atmosphere, were it not for the influence of the solar rays, which must, at least on those parts of the moon's surface exposed to the sun, counteract the effects of the cold in the body of this satellite, supposing such cold really to exist. There is, therefore, some difficulty in considering the moon to be totally devoid of gaseous matter on its surface, if volcanic action be still frequent upon it, as certain appearances would lead us to conjecture.

The white spots on the poles of Mars are, it is imagined, due to the presence of snow, "as they disappear when they have been long exposed to the sun, and are greatest when just emerging from the long night of their polar winter*." Now, if this opinion be founded on probability, it goes far to show that the inorganic matter on the surface of Mars may be similar to that on the surface of the Earth, and obeys the same laws.

It by no means follows, if even the planets be composed of somewhat similar materials, chemically considered, that all the bodies of our solar system are formed of similar matter. We have direct evidence to the contrary in the luminous envelope or atmosphere of the Sun, unless indeed we suppose some of the laws governing matter on the great scale to be brought, by design, into visible and intense action on the surface of the sun, for the benefit of the various bodies which hold their courses round it, while the same laws are in a feeble state of activity, or dormant, in these bodies. Be this as it may, it does not follow, because a body is wholly gaseous or composed of vapour, that intense heat necessarily exists in it. A highly elevated temperature is no more essential to the existence of that strange mass of light vapour, termed Encke's Comet, than to the existence of our atmosphere.

* Herschel, Treatise on Astronomy, p. 279.

Whether the materials of the planets were, or were not, originally, at a high temperature, the forms of these bodies seem to require the once free passage, among each other, of the particles of matter composing them. Now, this condition certainly does not exist at present on the surface of our earth, at least in those parts of the mineral crust which either protrude into the atmosphere or constitute the bed of the ocean. Hence a great change has taken place in the condition of our globe. We cannot, by any repetition of such effects as we now witness, conceive the possibility of a free passage of the particles of matter forming the earth among each other. When we suppose that seas can transport detritus from one distant place to another, and remove by their action any protruding portion of dry land, we assume the preexistence of solid matter, and, consequently, that the particles of such matter did not move freely among each other. To produce an action of water capable of wearing away solid matter, the water must either flow on land, be projected in the shape of waves on cliffs, or cut away the ground against, or on which, it may rest while running in streams or currents. It is clear, therefore, that solid matter must have existed before those rocks, commonly termed mechanical, could have been deposited; and, consequently, land either rising above the seas, or situated at a moderate depth beneath their levels, must also have existed, otherwise there would have been no action of water on the land sufficient to have formed detritus. Now the cubical contents of the mechanical rocks, known to us, is immense; so that, with every allowance for the production of siliceous and other minerals, from chemical solutions or combinations, during the deposit of the mechanical rocks, there remains a large amount of matter that must have existed in a solid state prior to the deposit of any mechanical rock. However true, therefore, it may be, that, if time sufficient be allowed, the action of water on land would tend to produce the spheroidal form of the surface of our planet, we are compelled to admit the prior solidity of land in such situations as would permit the removal of solid matter by the action of water upon it.

There is much difficulty in conceiving the earth to have revolved round the sun, as a planet, otherwise than with a spherical or spheroidal form. To assume that it was once an irregular solid, of a rough and uneven surface, and that it has been ground subsequently and externally by the action of water into a spheroid, seems but a clumsy hypothesis at best, and by no means accords with that simplicity which so preeminently distinguishes all the great works of creation. It would, moreover, but ill agree with the great mass of geological phænomena known to us. It seems fair to infer that, before the solid surface of our globe had been abraded by water on the one hand, or capable of supporting deposited detritus on the other, such surface was spherical or spheroidal.

When we regard the chemical composition of that part of the earth which falls under our examination, we are struck with the enormous volume of oxygen entering into the composition of all portions of it, whether air, water, or solid rock. Oxygen constitutes about 20 per cent. of the volume of the atmosphere; it forms a third part, by measure, of the gases composing pure water; and is locked up to an immense amount in the various rocks which are, viewed on the large scale, little else than a mass of oxidized substances. We shall not be far wrong if we estimate silica as constituting at least 45 per cent. of the mineral crust of our globe. Now silica is, according to Berzelius, compounded of 48·4 parts of silicium, and 51·6 parts of oxygen. Hence, if the oxygen confined in silica alone were thrown in its gaseous state into the atmosphere, the volume of the latter would be increased to an immense extent; and if the same substance were freed from the other mineral compounds forming solid rocks, and from the waters of the globe, the increase would be enormous.

The relative abundance of either hydrogen, nitrogen, carbon, sulphur, or chlorine, is by no means so considerable as that of oxygen. Hydrogen is known to us as entering into the composition of all waters, and as evolved in a compound state from volcanos, from certain fissures in the earth, and in districts where coal is found. It also enters into the composition of coal and similar mineral products. We must measure the volume of hydrogen principally by the amount of water, either present in the atmosphere, forming seas, lakes and rivers, or mechanically disseminated among rocks. As two volumes of hydrogen unite with one of oxygen in the production of water, it follows, as far as respects the water of our planet, that the volume of hydrogen is double that of oxygen. As the aqueous vapour disseminated in the atmosphere is continually varying, it is difficult to estimate its amount; indeed, water so circumstanced is little else than on its passage from one part to another of the solid or liquid surface beneath, so that it can scarcely be said to belong to the atmosphere, though the latter, to answer one of its great objects, is never without aqueous vapour.

As far as the superficies of our planet is concerned, water so predominates that, at first sight, hydrogen might be considered as constituting a substance of more relative abundance than it really does. Before we estimate the amount of hydrogen in the ocean, we have to deduct the amount of salts in solution. This is no large deduction, not amounting to more than 3 or 4 per cent. Regarding the area of the ocean, and taking its mean depth at about three miles, we certainly have, it must be confessed, an immense volume of hydrogen locked up. To this we must also add, as highly important, the great quantity of water mechanically disseminated through rocks. Much of this water is, no doubt, merely received from the atmosphere to be again discharged in the shape of springs, the rocks only acting as a reservoir for the time; and certainly a more beautiful or simple contrivance for the fertilization of the earth and the support of animal and vegetable life cannot well be imagined. There still, however, remains a quantity of concealed water, which in the aggregate must be very considerable, disseminated through rocks. Even in those rocks which merely supply springs, the amount of disseminated water must be enormous; for they so far resemble filters, that they are necessarily charged with the fluid before they permit it to pass out.

Capillary attraction must have great power, both in mechanically disseminating water among rocks, and in retaining it in them when so disseminated; it therefore keeps them, to a certain extent, saturated with moisture, and assists in promoting a more equal flow of water in springs. Capillary attraction and gravity no doubt carry water down far beyond those situations where it can be returned in springs, at least cold springs, for there are certain circumstances connected with those which are thermal, pointing to the conclusion that the water thrown up by them may have percolated to considerable depths. We may fairly consider that most rocks contain disseminated moisture, for there are very few which, when exposed to the proper heat, do not give out water. Some serpentines contain as much as from 12 to 15 per cent. of it. Water, moreover, enters into the composition of several minerals, apparently as a constituent part of them; but probably its amount, from this cause, is not of very great comparative importance.

The quantity of hydrogen locked up in coal and lignite would appear to be considerable. According to Dr. Thomson, cannel coal contains 21·56 per cent. of it. The same author states its amount at not more than 4·18 per cent. in Newcastle caking coal, so that it varies materially as a constituent part of this mineral. If it exists highly compressed, and even liquid, as there is some reason to believe, united with carbon, forming carburetted hydrogen, in the vesicles of coal, no small volume must occur in this state. In this account we must not neglect the hydrogen evolved from volcanos, either in aqueous vapour, or combined with other substances in a gaseous form. It would appear, however, that the evolution of hydrogen, either from volcanos, or from fissures in the rock, in the shape of inflammable gases*, produces little effect on the atmosphere, so that either the general amount must be inconsiderable, or it unites with the oxygen of the atmosphere, when in contact with the incandescent matter of volcanos, or exposed to an electrical discharge. Upon the whole, we may regard hydrogen as the next important substance of its class, which enters into the composition of the crust of our globe.

Nitrogen is chiefly important as constituting about 80 per cent. of the atmosphere. Its existence in animal and vegetable life may be considered as secondary, that is, derived from the atmosphere in the first instance. There is also every reason to suppose that it cannot be absent from numerous rocks which contain the organic remains of animals that have been entombed living, or at least with their flesh upon them. We have direct evidence in coal that nitrogen forms a portion of what may be considered solid rock. Dr. Thomson found it to constitute 15·96 per cent. of the Newcastle caking coal, and there is good reason for supposing that this is rather an under estimate of the amount of nitrogen contained in some coal, judging at least from the abundance of ammoniacal products given out from the distillation of coal in gas-works.

Carbon, independent of its existence in living vegetable and animal substances, is entombed to a large amount in fossil vegetables and in limestone. It is contained to the amount of 75·28 per cent. in Newcastle caking coal, of 75 per cent. in Glasgow splint

* See Geological Manual, 3rd edit., p. 151, &c.

coal, and of 64·72 per cent. in cannel coal. It is, however, in the mass of limestone, occurring on the earth's surface, that we should probably discover the principal amount of carbon. If carbonic acid be composed of equal volumes of the vapour of carbon and oxygen, the volume of the vapour of carbon condensed in the calcareous strata must be very great. Taking the specific gravity of pure limestone at 2·7, and estimating the weight of 100 cubic inches of carbonic acid at 47·377 grains, every cubic yard of pure limestone would contain 17092 cubic feet of carbonic acid gas*. As, however, limestone is rarely pure over considerable areas, we shall probably not be far wrong if we estimate the average amount of carbonic acid, locked up in every cubic yard of limestone, at about 16000 cubic feet. If, therefore, the carbonate of lime constituting not only the limestones, but also the calcareous matter disseminated through various rocks, were decomposed over the surface of the earth, the volume of carbonic acid set free would be immense.

The carbon in the atmosphere is not considerable, but without it vegetation could not exist. Theod. de Saussure ascertained that 10,000 parts of atmospheric air contained, as a mean, 4·9 of carbonic acid. There is a constant discharge of carbonic acid into the atmosphere from numerous springs and fissures in the earth†. M. Bischof estimates that 219,000,000

* This calculation is founded on the specific gravity of a very pure specimen of Carrara marble, sent me for the purpose by Mr. Chantrey. I found the specific gravity of an average specimen of carboniferous limestone from Bristol to be = 2·75, exceeding that of the Carrara marble by ·05.

† See Geological Manual, 3rd edit., *Art.* Gaseous Exhalations.

pounds of this gas are evolved from the vicinity of the lake of Laach in one year, which, taking 100 cubic inches of carbonic acid to weigh 47·377 grains, would give a volume of about 1,855,000,000 cubic feet of this gas, rising annually into the atmosphere from an area of a few square miles. These considerable evolutions of carbonic acid are merely local, and confined for the most part to districts of ancient or active volcanos; yet the collective amount of such discharges of gas from the surface of the earth into the atmosphere must be far from inconsiderable, more particularly when we take into account the mean annual volume thrown out from volcanic vents themselves.

Sulphur must be a more abundant substance in rocks than, at first sight, we might suppose. We are in the habit of associating our ideas of the connexion of sulphur and rocks more especially with volcanic products. When so estimated, the amount, though often locally great, is, taken generally, not considerable. Sulphur is, however, widely disseminated among many rocks. As a sulphuret of iron, it is distributed over the surface of the earth to a great extent, more particularly in those rocks which I have elsewhere termed the superior stratified or fossiliferous rocks*, and in those usually known as trappean. In many clays sulphuret of iron prevails to a great extent, and not a few organic remains are mineralized by it. Iron pyrites is a necessary and abundant substance in those shales whence alum is prepared, and thence named alum-shales. By far the largest por-

* See Geological Manual, *Art.* Classification of Rocks.

tion of the ores of copper and lead worked in different parts of the world are sulphurets. In the state also of sulphate of lime, sulphur is widely spread. Nothing is more common than selenite in beds of clay, and gypseous masses are sometimes of considerable magnitude. Not only does sulphur thus occur among rocks, but it is also disseminated throughout the ocean, sulphate of soda being one of the salts constantly present in all analyses of sea water. M. Eichwald states that sulphate of magnesia is a common salt in the waters of the Caspian Sea. We may therefore consider that sulphur is far from being a rare substance on the surface of the earth.

Chlorine is principally important as disseminated throughout the ocean, muriate of soda being the most abundant salt contained in sea water, and constituting about 2·5 per cent. of the whole saline solution. Muriate of magnesia and muriate of lime, other salts in sea water, though of minor importance, must still, collectively considered, contain a considerable amount of chlorine. No small volume of this substance must also be locked up in rock salt. When we regard the great quantity of matter which has been deposited mechanically, and sometimes chemically, in the sea, it is rather surprising that there are not greater indications of the presenee of the chlorides in them than chemists have as yet detected.

Of the other simple non-metallic substances, as they are termed, phosphorus, boron, selenium, iodine, bromine and fluorine, the relative geological importance is not very considerable. Phosphorus is principally known as entering into the chemical composition of animals. Human bones contain, according to Berzelius, 51·04 per cent. of phosphate of lime; and the enamel of teeth is stated by Mr. Pepys to be composed of 78 per cent. of the same substance. As a component part of minerals, phosphorus is rare; but there must be some amount of it entombed in various fossiliferous rocks, for the fossil bones of all ages contain it. Dr. Turner found 50 per cent. of phosphate of lime in a rib and a tooth of an Ichthyosaurus from the lias of Lyme Regis, and 29 per cent. in a vertebra of the same creature. He also detected 24·4 per cent. of phosphate of lime in a fish palate from the carboniferous limestone of Bristol, and 18·8 per cent. in a palate from the chalk*.

Boron enters into the composition of several minerals, but they are not such as constitute great masses of rock, with the exception of shorl rock, not unfrequent in some districts, such as Cornwall and Devon, at the skirts of the granitic masses, where the latter come into contact with the slates. According to calculation, boracic acid would form 1·79 per cent. of a shorl rock composed of equal parts of shorl and quartz. Boracic acid is also detected in some thermal springs. Selenium is only known in such small quantities as to possess little geological interest. Iodine is probably disseminated, though in exceedingly minute proportions, throughout the ocean, whence it is considered that sponges, sea-weeds and many marine creatures obtain it: it is also detected in many mineral springs. Bromine is probably also disseminated in all sea water; and Dr. Daubeny and other chemists have detected it in many mineral springs. It still

* Geological Manual, pp. 348 and 413.

further resembles iodine in its mode of occurrence, as it is found in the ashes of some sea-weeds and marine animals.

Fluorine is of more geological value than the simple substances immediately preceding, entering into the composition of some minerals which form constituent portions of great masses of rocks. Fluoric acid is found in mica and hornblende, two minerals of very great importance, particularly the former, as component parts of many rocks, the solid contents of which are, comparatively, very considerable. Fifteen analyses of mica, from various parts of the world, by Klaproth, Vauquelin, Rose and Beudant, afford, as a mean, 1·09 per cent. of fluoric acid; and Bonsdorf's analysis of hornblende, from Pargas, (a fair mean of various analyses of hornblende from different places,) gives 1·5 per cent. of the same substance. Calculation affords us 0·36 of fluoric acid in gneiss with mica, 0·54 per cent. of the same substance in mica slate, 0·75 per cent. in hornblende rock and greenstone, 0·18 to 0·21 per cent. in granite with mica, 0·5 per cent. in granite formed of quartz, felspar and hornblende (sienite), 0·65 per cent. in granite composed of quartz, felspar, mica and hornblende, and 0·5 per cent. in porphyritic greenstone*. Fluoric acid probably also enters, as a constituent part, into several trap rocks, difficult to class, but which there is every reason to conclude contain hornblende. If, therefore, we were certain that all micas contained fluoric acid, we might conclude that fluorine was not without a certain importance in the composition of the mineral

* Geological Manual, pp. 440, 441, and 449, 450.

crust of the earth; but we must not forget that lithia takes the place of fluoric acid in certain micas. To what extent lithia micas prevail is not exactly known; but there is always a possibility that, though their relative amount may now be considered small, more exact examination may find it large. Fluor spar is, doubtless, the mineral in which the greatest relative amount of fluorine is detected, but, geologically considered, it is one of little importance.

Of the metallic bases of the alkalies and earths, silicium is the most abundant on the surface of our planet, silica entering so largely into the composition both of the chemical and mechanical rocks. According to calculation*, silica is contained in the rocks enumerated beneath in the following proportions:

	per cent.
Gneiss	70·06 to 71·86
Mica slate	61·94 to 73·07
Hornblende rock	54·86
Chlorite slate	63·71
Talcose slate	78·15
Compact felspar	51·00 to 60·00
Granite	63·96 to 74·84
Schorl rock	68·01
Greenstone	54·86
Hypersthene rock	59·14 to 61·85
Basalt	44·50 to 59·50
Pitchstone	72·80 to 73·00
Serpentine	42·00 to 43·07
Diallage rock	58·42 to 60·55

In pure quartz rock silica would be the sole ingre-

* Geological Manual, p. 440—443, and p. 449—454.

dient. When quartz rock is composed of equal parts of quartz and felspar, silica would form 82 per cent. of the constituent parts. It is also abundant in the rocks of decidedly mechanical origin. The greater part of the immense mass of conglomerates, sandstones and slates known as grauwacke, is composed of silica. The same with the old red sandstone (if it be considered a separate rock from grauwacke), the coal measures, the various rocks known as the new red sandstone, numerous beds of sandstone and clay in the oolitic group, the various sands and sandstones of the Wealden rocks and of the cretaceous group, and numerous rocks of the supracretaceous group. It is often disseminated among the calcareous beds themselves, even to a considerable amount. This is well seen in portions of the chalk where the flints constitute nearly one third of the whole mass. Many other limestones contain silica, as the lime-burners sometimes find to their cost, the lime and the silica forming, from the heat of the kiln, silicate of lime.

Next to silicium, aluminium would appear to be the most important base of the earths on the face of the globe. Its collective amount is by no means so great as that of silicium, but it is quite as widely spread. Even limestones are rarely so pure as to be without it; many indeed contain it in considerable quantity, and are valuable for water-setting purposes in consequence. By calculation*, alumina would exist in the rocks enumerated beneath in the following proportions:

* Geological Manual, *Art.* Inferior Stratified Rocks, and Unstratified Rocks.

	per cent.
Gneiss	15·20
Mica slate	13·08 to 15·45
Hornblende rock	15·56
Chlorite slate	8·95
Talcose slate	13·20
Compact felspar	15·00 to 30·00
Granite	10·37 to 14·32
Schorl rock	17·91
Greenstone	15·56
Hypersthene rock	10·59
Basalt	11·50 to 16·75
Pitchstone	10·84 to 11·50
Diallage rock	13·14 to 13·86

There is scarcely any among the mechanical rocks that does not contain alumina. It constitutes the base, as is well known, of the various clays, and must be regarded as a very abundant and important constituent part of rocks.

Potassium and sodium appear to be the next most important metals of their class. Potash is more abundant than soda as an ingredient of rocks, and can only, notwithstanding its name, be considered as existing in a secondary condition in vegetation, that is, the potash of plants is derived from the decomposition of rocks containing that substance. Though very widely disseminated, its amount, collectively considered, is very inferior to that of silicium and aluminium. There are few, if any, of the inferior stratified rocks without potash; and it may be stated, as a fair approximation to the truth, that it constitutes from 5 to 6 per cent. of such rocks, viewing them in the mass. It is more or less abundant, according to circumstances, in almost all the decidedly detrital rocks. Indeed, we can scarcely consider it otherwise than present in the greater proportion of mineral masses. It may be calculated that potash constitutes between 6 and 7 per cent. of the mass of granites, and about 7 per cent. of greenstone and rocks of that class.

Sodium is chiefly important in rocks from its presence in certain felspars, thence named soda felspars: these constitute a component part of certain of the gneiss family, and of granites. Soda is found in schorl and certain hypersthene rocks, in some eurites, in trachytes, pitchstones, basalts, and some diallage rocks. Beudant found that it amounted to 5·9 per cent. in a basalt from Baulieu. There can be little doubt that a large amount of soda is thus locked up in rocks, particularly when we include the masses of rock salt, discovered in different parts of the world; but it would appear to be most widely desseminated in the ocean, constituting an essential part of the principal salt dissolved in its waters.

Magnesium and calcium are the substances which appear to succeed in importance: of the two, magnesium is probably the most abundant. Magnesia is present in all the inferior stratified rocks, with the exception of quartz rock (without mica) and certain eurites or compact felspars. In the detrital rocks it is also common, particularly when mica forms any considerable portion of them. There are few limestones which do not contain magnesia: in some it is abundant, as in those termed magnesian limestones. It is an essential ingredient of dolomite, properly so called, carbonate of magnesia constituting more than

40 per cent. of that rock. Magnesia is also disseminated through the waters of the ocean, muriate of magnesia forming from ·004 to ·005 of their mass.

The relative value of calcium is more apparent in the more modern than in the more ancient rocks; for although it is always found in gneiss, mica slate, chlorite slate, talcose slate, clay slate, eurite, and hornblende rock, it is always in very small quantities, with the exception of the last, in which it amounts to 7·29 per cent.; taking the others as a whole, it would not form more than ·55 per cent. of the mass. In this account we have neglected the limestones and dolomites associated with the inferior stratified rocks. They form a very small portion of such rocks, and would not add greatly to their collective amount of lime; but they are important, as the lime exists as a carbonate in them, while in the others it is in the condition of a silicate. Lime is found in all granites, and with the exception of the granite with hornblende, in very small quantities. It is also discovered in greenstone and all trappean rocks, in hypersthene rock, basalt, pitchstone, serpentine, diallage rock, and trachyte. There has been occasion elsewhere to remark*, that the relative fusibility of the igneous rocks depends upon the amount of silicate of lime found in them, and therefore that rocks containing much hornblende or augite are more readily acted on by heat than the others. Lime is more abundant among the fossiliferous rocks, particularly towards the central and higher part of the series, where it chiefly exists as a carbonate. Lime is also disseminated through sea waters, though in small quantities; so that calcium is widely distributed in land and water,

* Geological Manual, p. 454.

being principally abundant in the central and higher parts of the fossiliferous rocks, and widely dispersed in small quantities throughout the more ancient rocks, and in the waters of the ocean.

The other bases of the alkalies or earths, barium, strontium, glucinium, yttrium, thorium, zirconium and lithium, occur in quantities too small to be of geological importance, with the exception perhaps of lithium, which may be contained in more micas than is generally supposed.

Of all the metals, the oxides of which are neither alkalies nor earths, iron and manganese are the most important, geologically considered. Calculating the mean of thirty different kinds of rock, and neglecting iron ores, properly so called, of every kind, iron constitutes, as an oxide, 5·5 of the lowest stratified rocks, amounting to 14·72 per cent. in mica slate with garnets, and 15·31 per cent. in chlorite slate. It forms 12·62 per cent. in hypersthene rock, and about 20 per cent in basalts. Oxide of iron constitutes between 2 and 3 per cent. of the mass of granites and gneiss, and between 3 and 4 per cent. of the mass of greenstone and the more common trappean rocks. When we consider the large amount of iron which exists either in the state of an oxide, a carbonate, a carburet, a silicate, or a sulphuret, therefore including all iron ores of importance, and also regard the proportions which the various rocks bear to each other, we shall probably not err greatly if we estimate iron as constituting about 2 per cent. of the whole mineral crust of our globe. It is remarkable that manganese is almost, though not quite, as widely disseminated through rocks as iron, the proportions in which it enters into their composition being, however, much smaller. There is scarcely a rock without iron, and there are very few which do not afford some trace of manganese; except, however, in the places where its ores are worked, the latter exists only in minute quantities. Its calculated amount is greatest in mica slate with garnets, where, as an oxide, it forms 1·23 per cent. of the constituent parts of the rock. Taking rocks generally, manganese cannot be estimated as forming more than ·03 or ·04 per cent. of the mass.

The other metals, such as tin, copper, lead, zinc, arsenic, silver, gold, &c., are, when we view rocks in the mass, of little geological importance: they are principally found in veins. When regarded with reference to their presence in such veins, and their general mode of occurrence, they possess very high geological interest, but their relative amount among the substances composing the crust of the earth must be inconsiderable. Chromium is so frequently discovered among rocks of the serpentine family, that there may be some connexion between them, and it is possible that the colour of these rocks may partake of the green tint of the oxide of chromium. Titanium may also be more widely disseminated than is supposed, for it is a common accompaniment of iron ores. The manner in which it is discovered in some, is a strong proof that metals may be widely disseminated, and yet escape detection, unless large masses of the substances containing them be subjected to chemical action*.

The principal substances entering into the chemical composition of our planet's surface may be classed in

* Minute crystals of titanium are discovered in the slag from large iron works, more particularly from those of Merthyr Tydvil.

the following order, according to the respective importance of each.

Simple non-metallic substances.

1. Oxygen.	3. Nitrogen.	5. Sulphur.	7. Fluorine
2. Hydrogen.	4. Carbon.	6. Chlorine.	8. Phosphorus.

Metallic bases of the alkalies and earths.

1. Silicium.	3. Potassium.	5. Magnesium.
2. Aluminium.	4. Sodium.	6. Calcium.

Metals, the oxides of which are neither earths nor alkalies.

1. Iron.	2. Manganese.

It would therefore appear that sixteen substances, commonly considered simple, constitute by their various combinations, if not entirely, at least by far the largest amount of, all the matter which, either gaseous, liquid, or solid, organic or inorganic, is known to exist on the surface of the earth.

CHAPTER II.

There is so much grandeur and simplicity in the idea of the condensation of gaseous matter into those spheres or spheroids which exist, not only in our solar system, but also by myriads throughout the universe, that we are irresistibly led to adopt some view of this kind, more particularly as it would accord with the unity of design so evident throughout creation. Encke's comet, that remarkable body of vapour which revolves round the sun in about 3¼ years, proves, by its existence, that gaseous matter or vapour, of extraordinary tenuity, may float around our great luminary in given times, and in a given orbit, checked only by a resisting medium of still more extraordinary tenuity. There is therefore no argument, *à priori*, against the hypothesis that the matter composing our globe may once have existed in a gaseous state, and in that state have revolved round the sun. We might even go further, and consider, with La Place, that our whole solar system is but a condensation into parts, doubtless from design, of that matter which now constitutes the sun, the planets and their satellites,—matter which rotated on an axis, and hence the fact that all the planets move in the same direction. In support of this view, let any one weigh the evidence recently adduced respecting nebulæ, more particularly by Sir John Herschel*, and he will have some difficulty in resisting the impression that these bodies are enormous masses of

* Phil. Trans., 1833, and Treatise on Astronomy.

matter in the act of condensation. If all the matter existing in the sun, planets and satellites were expanded to, and even beyond, the orbit of Uranus, the whole mass would still be but a speck in the universe.

Heat would act the part of an antagonist force to the condensation of the matter composing our globe, supposing it to exist in a gaseous state. It would not only tend to resist the action of gravity, but also the attraction of chemical affinity. Now, if it be highly probable, as has been previously observed, that heat to a certain extent resists the action of gravitation in the sun, Jupiter, and Saturn, there is nothing unphilosophical in the inference that heat has resisted, and may continue to resist, in a minor degree, the action of gravitation in our planet. We shall have occasion to show, from other reasons, that the latter inference is highly probable, and if so, from the nature of things, a greater and a previous resistance is not less so.

The probable effects resulting from a mixture of all terrestrial matter in the state of gas or vapour it would be exceedingly difficult to appreciate, inasmuch as we are unacquainted with the matter beneath the crust of the globe; and even if we were, the necessary calculations would be so intricate, that it is extremely doubtful if our actual knowledge could carry us to the end desired. Let us, however, for a moment consider what might be some of the effects of diminished temperature on the principal substances constituting the earth's crust, including the seas, lakes, rivers and atmosphere, and see how far the calculated effect might accord with the present distribution and condition of such matter.

So long as matter exists in the state of gas or

vapour, there is reason to conclude that the different kinds would be permeable to each other; at least experiments on gases would lead to this inference. Hence, supposing, for the sake of the argument, that the heat were sufficiently intense, the simple non-metallic substances, and the vapours of the various metals, would tend to mix with each other. This condition of things would not continue to the external part of the sphere or spheroid, the existence of which we now suppose; for the temperature would become less, from various obvious causes, at the outer parts, and the vapours of a great proportion of the metals would cease, from want of the necessary heat, to exist. They would tend to condense and to separate from the mass of the non-metallic simple substances, neglecting for the moment any chemical affinity which may exist between the metals and certain of those substances. A condensation of the particles of metallic vapours would cause them to lose their support among the particles of gaseous matter, and the action of gravity would tend to carry them towards the centre of the sphere; but as they could not pass beneath the point where the heat would again convert them into vapours, we should obtain an inner sphere or spheroid of metallic vapours, striving to condense, surmounted by a body of the non-metallic simple substances, which could readily exist, some even to the extreme superficies of the whole sphere or spheroid, at a greatly inferior temperature*. We must

* It is not intended to infer from this view that the non-metallic simple substances could not exist beneath, or intermingled with the inner sphere or spheroid; but merely that a mass of such substance rose above it externally.

not here neglect the action of gravity. It has been assumed, that the heat being sufficient to counteract this action to a certain amount, all terrestrial matter was gaseous. The struggle between these antagonist forces would be most powerful, for as the volume of gaseous fluids is inversely as the pressure to which they are exposed, the pressure upon the internal portions of the gaseous sphere or spheroid would be enormous, and therefore, when, from that radiation of heat which must take place into the cold planetary spaces, gravity came forcibly into action, liquids and solids would necessarily result from this cause alone, and particles of matter be squeezed together, even into liquids and solids, in the interior, which would retain a gaseous form on the surface at the same or higher temperatures.

If we consider that the attraction of chlorine for metals is greater than that of oxygen, it does not at first sight appear probable that when a diminished temperature permitted the metals to become either liquid or solid, they should unite with oxygen, (as we find has been the case in the mass of the older rocks,) in preference to chlorine, when both were equally present, independently of any combinations formed when the metals were in a state of vapour. As a constituent part of the globe's surface, the volume of chlorine is greatly inferior to that of oxygen, if both only existed in a gaseous state.

It is considered that chlorine and hydrogen, in the dark and at common temperatures, may remain together, as gases, for any length of time without combining, notwithstanding the powerful affinity of the former for the latter, which is so great that chlorine will decompose water by the action of light, to unite with hydrogen. In the case we have supposed these substances would neither be in the dark, nor exposed to low temperatures. They would, however, not be alone together; they would be mixed with other simple non-metallic bodies and with metallic vapours. Now if chlorine should unite, in consequence of its great attraction for metals, with their vapours, the union would probably not continue after the formation of water or aqueous vapour. For, if the most abundant metals should have a strong affinity for oxygen, as silicium, aluminium, potassium, sodium, magnesium and calcium have, the water or aqueous vapour would probably be decomposed, the hydrogen going over to the chlorine, and the oxygen to the metal. Thus a large amount of the chlorides would disappear, and oxides would be abundant, independently of the direct formation of the latter by the union of oxygen with the metallic bases of the earths and alkalies. The union of the chlorine and hydrogen would form muriatic acid gas. This combination once formed, the value of chlorine over oxygen in its attraction for metals is destroyed, and the two latter can freely unite, as they have done, in the production of great masses of rock, which are, as has been previously noticed, little else than metallic oxides.

The absence of the chlorides from rock masses, (if we except rock salt, the production of which must be considered as resulting from secondary agency,) is very remarkable, and agrees with the above hypothesis.

Hydrogen unites with few metals, and among them potassium is the only one of importance to our present subject. It, however, possesses a strong attraction for oxygen. Although it does not unite with oxygen, even when mixed in the proper proportion of two to one, at ordinary temperatures, except by the electric spark, by flame, or a body heated to bright redness, it nevertheless does quietly combine with that substance, when they are exposed to a temperature above the boiling point of mercury, and below that at which glass begins to appear luminous in the dark. Hydrogen also silently unites with oxygen, by a succession of electric sparks, when diluted with too much air to explode by the same means; and spongy platinum causes hydrogen to combine slowly with oxygen, even when mixed with one hundred times their bulk of oxygen gas*. The power of oxygen and hydrogen gases to unite quietly, even when mixed beyond the proportions in which they exist in water, without the necessity of explosion, is an important consideration in our present inquiry, because it seems to point out the possibility of producing water by slow means when a large amount of oxygen, a less proportion of hydrogen, and a still less amount of nitrogen were mixed together, with minor proportions of a few other substances; in fact, under conditions which might exist in the exterior parts of the gaseous sphere or spheroid under consideration.

If water or aqueous vapour were produced, either quietly by some such means as those above noticed, or more suddenly by explosions, it would be speedily seized upon by the muriatic acid gas, from the well known powerful affinity of the latter for water; an attraction so strong, that the mere escape of this gas into common air causes a dense white cloud, from its

* Turner's Elements of Chemistry, 4th edit., p. 219.

immediate combination with the aqueous vapour in the atmosphere.

When we view the present distribution of the principal substances constituting the earth's surface, including the ocean, we are struck with the fact, that while sodium exists both in rock and water, it is principally as a chloride in the latter, while it is united with oxygen in the former. Indeed, the mass of chlorine is combined with sodium and dissolved in the waters of the ocean, for rock salt is a secondary production, not found in the oldest stratified rocks. Chloride of sodium is also not uncommon in mineral springs. The chlorine of marine vegetation can only be considered in a secondary light, and cannot enter into our present calculations. It does, therefore, seem probable that there was once a condition of the globe when chlorine by combining with hydrogen and sodium, and hydrogen by uniting with oxygen, laid the foundation of the actual ocean, omitting for the present the consideration of the muriates of magnesia and lime, which, with sulphate of soda, also exist in minor proportions in the waters of the sea. It is worthy of observation, that however hypothetical the gaseous condition of our planet may be, it does not directly oppose any serious obstacle to the production of an ocean in which the great amount both of hydrogen and chlorine should be combined; for even supposing that, from intense heat, such combination would not in the first instance be absolutely liquid, it must necessarily become so when the radiation of heat was sufficiently advanced to permit the existence of a moderately cool solid beneath it. By this hypothesis we at once dispose of the chlorine and hydrogen; for the latter, like the former, only exists in a secondary condition among rocks.

Nitrogen presents little difficulty, It seems principally designed to correct an excess of oxygen in the atmosphere, and to enter into the composition of certain of the organic substances which exist in it. Hence, probably, the reason why nitrogen does not constitute a portion of masses of rock, except where they contain organic remains. From the beginning, nitrogen, viewed in the large scale, would probably not combine to any extent with liquids or solids, until the atmosphere was finally fitted for the existence of animal and vegetable life. When organic exuviæ were entombed and became portions of rocks, nitrogen also entered into the composition of the mineral strata. It does not occur as the component part of the older rocks.

Carbon is so rare among the older rocks, that under whatever conditions they have been produced, some bar to its union with the lime, magnesia, potash, soda and other substances contained in them must have existed, even when it was combined with oxygen in the shape of carbonic acid. Carbon would, under the conditions we have supposed, most probably combine with oxygen, and produce this acid. Now the absence or, rather, comparative rarity of the carbonates among the older strata, and the prevalence of the silicates in the same rocks, seem to point to the prevalence of considerable heat at the time when the minerals composing them were *first* formed; for though the carbonates can be readily fused beneath pressure, the necessary force must first exist, and the state of things we have been contemplating would scarcely permit this, at least at the beginning, or on the outer portions of the sphere. Supposing an union of carbon and oxygen in the proper proportions, and carbonic acid the result, being gaseous, it would readily permeate the other gases and vapours, and would have a tendency to float, mixed with them, in the higher or outer parts of the sphere or spheroid, while, from the radiation of heat, consolidation was proceeding in the inner portions, so that it might eventually constitute a considerable part of an atmosphere principally composed of nitrogen and oxygen. As it is no part of our intention to conceal difficulties, it must be stated that to suppose carbonic acid originally and in a great measure confined to a gaseous envelope of our planet, does not well accord with the production of limestones, nor with the evolution of this gas from volcanos, fissures in the earth, and from springs. It would, indeed, accord with the views of M. Adolphe Brongniart, who, to account for the early vegetation of our planet, considers the atmosphere to have once been more impregnated with carbonic acid than at present. That this has been the case is highly probable, more particularly when we regard the amount of carbon entombed in coal and fossil plants, which, from analogy, we consider to have been principally derived from the atmosphere. To account, however, for various geological phænomena, we require carbon as well beneath the crust of the earth as on its surface.

Sulphur is so readily volatilized, that it would be kept in the state of vapour, as far as regards mere temperature, after a considerable radiation of heat; but there are other conditions that must be considered. The vapour of sulphur would endeavour to combine with oxygen, and the result would be sulphurous acid; and this again, if aqueous vapour be present, would become sulphuric acid. The sulphates are not collectively abundant in rocks; they are unknown among the inferior stratified rocks, though sulphate of lime is not rare among the fossiliferous strata. The vapour of sulphur would also strive to unite with the metallic vapours and produce sulphurets; but so long as the heat was great, and oxygen present, this would be a difficult combination, for there is a tendency in all sulphurets to decompose under such conditions, however refractory, in this respect, some sulphurets may be when exposed to mere heat. The principal sulphurets contained in the older rocks are iron pyrites.

With respect to fluorine and phosphorus, the former has a powerful affinity for hydrogen; and we can scarcely consider it to exist without endeavouring to combine, when in contact with it. Phosphorus is so little important as a constituent portion of land, air, or water, except as a component part of some of the organic exuviæ entombed in the fossiliferous rocks, that we might disregard it in a general view like the present, did it not constitute so important a part of the bones of animals; and hence it is not a little interesting to inquire as to the manner in which it probably existed prior to the period when it was required as a portion of organized life. This does not readily suggest itself; and it is merely noticed for the purpose of promoting investigation.

If there be any approximation to truth in supposing that the metallic vapours would be arrested, by diminished temperature, at a certain height above the centre of the gaseous sphere or spheroid, and an inner spherical or spheroidal crust the result, the oxygen on the upper surface of the latter would combine more readily with the metals than on the under surface, from the difference of temperature. The oxides would tend sooner to consolidate above from the same cause, and to exclude the metals beneath from the action of the oxygen, which, from the high temperature of the interior, would be far more dense above than towards the interior, where, indeed, even supposing the metallic vapours to mix with the other substances as gases do, the oxygen would be of extraordinary tenuity. Hence, there might eventually be, from the radiation of heat, an oxidized solid crust surmounted by a gaseous envelope, in which oxygen was present, covering a still heated interior, composed of metallic substances but slightly mixed with oxygen and other simple non-metallic substances. This hypothesis would therefore produce a condition of things that would accord with the theories both of central heat and of the existence of the metallic bases of various substances beneath the crust of the globe, the oxidation of which may now produce many geological phænomena.

We have been induced thus to enter at some length upon the hypothesis of a sphere or spheroid, composed of the matter of the earth rendered gaseous by intense heat, solely to promote further inquiry. Fully to discuss the subject would be most difficult. It would require very complicated calculations respecting the action of gravity, heat, electricity and chemical affinity under such conditions; but if, finally, anything like a fair approximation to the present state of the earth's surface could be deduced from such calculations, a most important advance will have been made in theoretical geology.

Although a free passage of the particles of terrestrial matter among each other seems necessary to the production of the earth's figure, (an object of the first geological importance, however convenient it may be for those theories to neglect it to which it is fatal,) it by no means follows that this matter should have been altogether in a gaseous state in the first instance. A gaseous envelope, and a liquid metallic nucleus, would produce the effects required equally well, particularly if we suppose an internal heat so intense as to prevent the oxidation of the metallic nucleus except on the surface. An oxidized crust would bar the progress of oxidation except through cracks and fissures, and thus many geological phænomena would be explained. In fact, we should still have a condition of things favourable to an union of the theories of central heat, and of the chemical action of oxygen upon the metallic bases of certain earths and alkalies. The mean density of the earth being about double that of the oxidized mineral crust, it appears too little for the condensation of the same kind of oxides by pressure, from gravity, towards the centre. We may therefore consider the interior of the earth to be heated, if metallic, in order to account for the density.

It has been opposed to this view, that potassium and sodium, important substances in the earth's crust, are too light to produce the effect required, the density of potassium being only 0·865, that of sodium 0·972. Now, if we admit, for the sake of the argument, that all metals are capable of uniting with each other in certain proportions, producing numerous alloys, it does not follow that they can do so at all temperatures; the process of obtaining pure silver by amalgamation is a well-known example to the contrary, for though it depends upon the affinity of silver for mercury in the first instance, this would be entirely useless if the mercury could not be readily volatilized by heat, and thus leave the silver. Hence, we may not only conclude that the more volatile metals, such as potassium and sodium, would be driven to the outer portions of the metallic sphere or spheroid by great heat, but that they would have much difficulty in forming alloys if the temperature be very elevated, and from this cause also be driven to the surface, where a large portion of them would readily become oxidized, and possess the density of such compounds. Consequently, the small density of such metals as potassium and sodium would have little weight as an argument against the supposition of a metallic nucleus in our planet, one in which an elevated temperature still prevails, thus rendering the mass of comparatively small density, not exceeding the mean usually assigned it.

As we have elsewhere * accumulated the various facts observable on the earth's surface, which render such a central heat highly probable, it would be superfluous to adduce them here: it will therefore be sufficient to state the general results. 1. Numerous experiments in mines show, notwithstanding their liability to error from various causes, an increase of temperature from the surface downwards, that is, from those depths where the action of the solar rays ceases

* Geological Manual, p. 6—27.

to produce a variable heat, which, taking the experiments made in the vaults beneath the Observatory of Paris as a guide, would be from 60 to 80 feet. 2. Thermal springs occur in all parts of the world, and among all varieties of rock. They are generally discharged on the surface through fissures or cracks, produced by disrupting causes at various geological periods. 3. The temperature of the water in those perforations into the earth's surface named *Artesian Wells*, is found to increase with the depth, whence it occurred to M. Arago that the temperature of the earth at different depths might be ascertained by them. 4. Terrestrial temperature, at small depths, does not coincide with the mean temperature of the atmosphere above it. Wahlenberg remarks, and Kupffer confirms the observation, that many deep-rooted plants only flourish in the northern regions, because the mean temperature of the earth exceeds the mean temperature of the air in such situations. The experiments of Von Buch, Humboldt, Hamilton, Hunter, Smith and Ferrier tend to show that the temperature of common springs in the tropics is beneath the mean of that of the air in the same places. 5. Igneous matter has been ejected at all periods from the interior of the earth. 6. Active volcanos occur widely spread over the surface of the world, and present, in their general phænomena, so close a resemblance to each other, that they may be considered as produced by a common cause, and that cause deep-seated. 7. Geological phænomena attest a great decrease of temperature on the surface of the globe. 8. The water of seas and lakes merely arranges itself according to its greatest specific gravity, and therefore, in general, would afford little evidence in favour of, or against, central heat. There is, however, an increase of temperature with the depth in many places in high latitudes which does not accord with the arrangement of water according to this law. It may be assumed that such increase of temperature, interfering with the densities, is due to heat in the bottom beneath, which, though sufficient to produce a visible effect upon waters so closely approaching their maximum density as those in high latitudes, is insufficient to be apparent beneath warm climates. 9. A decreased temperature of the earth would, by radiation, and in accordance with the views of M. Elie de Beaumont, produce the various mountain ranges and fractured strata found on the surface of our planet.

When all these circumstances are taken into consideration, and we add the probability that heat counteracts the effects of gravity in the sun and certain planets, and that the free passage of the particles of terrestrial matter among each other was necessary to produce the figure of the earth, the evidence in favour, not only of a central heat at present, but also of a heat of far greater intensity at remote geological epochs, becomes exceedingly strong; so strong, indeed, that there is some difficulty in resisting the impression that we have, by various means, made as fair an approximation towards the truth, as the nature of the subject will admit.

It may here be remarked, as not a little curious, that geologists who are unwilling to admit the probability of a central or internal heat, still call in the aid of very intense heat, acting very generally over the earth's crust, to account for various geological phænomena. Some go so far as to consider that all the inferior stratified rocks have been deposits precisely similar to those now going on, even to have contained organic remains, and that their present character is due to the agency of great heat, which produced the necessary alteration. Without stopping to notice the physical and chemical objections to this view, it may be observed that the inferior stratified rocks being widely spread, with common characters, over the face of the globe, this hypothesis requires as high a temperature, and a heat as widely diffused, as the advocates for a central heat could possibly desire; a temperature, indeed, which, if this process be still considered in full force, would exceed that which their calculations would afford them.

While on the subject of the early condition of our planet, it is not a little interesting to observe the effects of pressure upon gases, as shown by Mr. Faraday*. The following table exhibits the pressure and temperature at which the gaseous substances enumerated beneath became liquid in his experiments:

Sulphurous acid gas.........	2 atmospheres at	45° F.
Cyanogen gas	3·6—————	45°
Chlorine gas.................	4 —————	60°
Ammoniacal gas	6·5—————	50°
Sulphuretted hydrogen gas.	17 —————	50°
Carbonic acid gas...........	36 —————	32°
Muriatic acid gas...........	40 —————	50°
Nitrous oxide gas...........	50 —————	45°

It is impossible not to be struck with the fact that chlorine gas is liquefied by so small a pressure as that of four atmospheres at a temperature of 60°, while

* Phil. Trans. 1823.

muriatic acid gas requires a weight equal to forty atmospheres to produce the same effect at 50°. So that while oxygen and hydrogen gases resist all the attempts which have been yet made to force them into a liquid state, chlorine gas requires but a small pressure for its liquefaction. This alone is an important element in calculations respecting such conditions of terrestrial matter as we have above supposed. It is no less interesting, that sulphurous acid gas becomes liquid under the very small pressure of two atmospheres, and that a compound of nitrogen and carbon, constituting cyanogen gas, is reduced to the same state by a compressing force equal to 3·6 atmospheres. Thus, though oxygen gas alone cannot by any means yet employed be compressed into a liquid, it, when united with an equal volume of the vapour of sulphur, merely requires a pressure equal to two atmospheres, at a moderate temperature (45° Fahr.), to produce this effect; and nitrogen, not yet rendered liquid by pressure in its simple state, readily takes that form, beneath a small weight, when united with twice its volume of the vapour of carbon. Again, hydrogen gas, which in its simple state has resisted all attempts to compress it into a liquid, is no sooner united with nitrogen, in the proportion of three measures of the former to one of the latter, producing two measures of ammoniacal gas, than the small pressure of 6·5 atmospheres, at the moderate temperature of 50° Fahr., will force it into a liquid state. Now this latter fact is very remarkable, for it shows that while a large proportion of hydrogen, by its union with a minor proportion of another simple non-metallic and, in the respect under consideration, highly refractory substance, produces a readily liquefied compound, an union of equal volumes of hydrogen and chlorine, the latter the only simple substance of its kind easily rendered liquid by pressure, produces a compound requiring forty atmospheres for its liquefaction. We must not here forget that water, the union of two gases which resist all our attempts to condense them by pressure, is liquid beneath a weight still less than that of one atmosphere.

Now it is important, in a geological point of view, to remember, that a very moderate increase of temperature causes gases liquefied by pressure to boil, and that the removal of the necessary compressing force produces violent explosions in some; so that if considerable heat should act upon certain of these liquid gases, carbonic acid, for instance, entombed in an uncombined state among rocks at moderate depths under the surface of the earth, the superincumbent matter would be rent and fractured, if the resistance above were unequal to the expansive force of the heated gas beneath.

To produce the liquefaction of the gases enumerated in the foregoing table would require, even for nitrous oxide gas, no great depth of rock. To show this more clearly, let us estimate the depth of water (neglecting the combinations of the gases with it,) necessary for the same purpose, taking the above experiments to be, if not absolutely true, very close approximations to truth. Estimating the height of a column of water equal to the pressure of an atmosphere, in the usual way, at 34 feet, and neglecting the saline contents of the sea, and the compression of the water itself at great depths, both of which would cause the same effects to be produced at less depths, we find that

Sulphurous acid gas could not exist beneath	68 feet,	at 45° F.
Cyanogen gas	123 —	— 45°
Chlorine gas...............................	136 —	— 60°
Ammoniacal gas	221 —	— 50°
Sulphuretted hydrogen gas..............	578 —	— 50°
Muriatic acid gas	1360 —	— 50°
Nitrous oxide gas	1700 —	— 45°

As carbonic acid gas requires a pressure of 36 atmospheres, at 32° F., for liquefaction, it cannot enter in this table, which supposes the water to be fresh, and consequently solid at 32° F. A greater amount of pressure would of course cause this gas to become liquid at a more elevated temperature; but we have every reason to conclude that deep masses of fresh water always remain at a temperature between 39° and 40°, that at which such water is most dense, so that we may suppose carbonic acid would remain gaseous, as far as mere pressure was concerned, beneath all the known fresh-water lakes of the world, their depths being insufficient to produce the effect required: it is otherwise, however, with the waters of the ocean. According to M. Erman, salt water of the specific gravity of 1·027 diminishes in bulk down to 25° F., and does not reach its maximum density before congelation. Dr. Marcet's experiments also point to a similar result, 22° F. being, according to this author, the temperature at which sea water reaches its maximum density. Carbonic acid gas would therefore become liquid, as far as regards pressure, beneath 204 fathoms of sea water, the temperature of which was 32° F. These conditions may readily exist in high latitudes; indeed, according to Capt. Ross, he found, in lat. 60° 44′ N., and long. 59° 20′ W., a temperature of 29° at 200 fathoms, 28° at 400 fathoms, and 25° at 660 fathoms. Other observers, however, give a more elevated temperature for the deep waters of high northern latitudes.

If we estimate the mean density of the mineral crust of the globe at 2·5, which is probably less than the truth, we find that sulphurous acid, cyanogen, chlorine, ammoniacal, and sulphuretted hydrogen gases would be squeezed into liquids beneath 250 feet of rock, while the same effect would be produced upon muriatic acid gas at 570 feet, upon nitrous oxide gas at about 750 feet, and upon carbonic acid gas at 1300 feet, allowance being made in each case for the increased temperature found at increased depths. These are important considerations, which not only have weight when we consider the early conditions of our planet, but also when we regard the chemical changes now in progress beneath the immediate surface of the earth, and which are manifested by various discharges of gaseous matter into the atmosphere.

CHAPTER III.

As there is great difficulty of conceiving any other than the igneous fluidity of our planet previous to the consolidation of its surface, and as fluidity seems essential to the figure of the earth, it has been suggested that the earliest transition from a liquid to a solid state, consequent on the radiation of heat, would take place at the equator, and that masses of the solidified crust would float upon the incandescent fluid beneath*. The fluid mass would necessarily be influenced by tides; therefore, so long as the solidified crust was too thin to resist the effects of this cause, it would be broken up into fragments, the precursors of those of which the solid surface of our globe is everywhere composed. It may surprise many of our readers, but it is nevertheless true, that when the surface of England is minutely examined, we find there is scarcely any area of eight or ten square miles in extent which has not been fractured, or broken up into minor portions, by causes that have acted at various geological epochs. What is thus true of England is found to be also generally true of the whole surface of our continents and islands when examined with the necessary attention. Sometimes a more modern deposit may mask a surface broken into fragments, the former not having been yet acted on by disrupt-

* Croizet et Jobert, Recherches sur les Ossemens Fossiles du Département du Puy-de-Dome, 1828.

ing forces; but when rocks, on which such deposits rest, are exposed by denudation of any kind, either in ravines or over a certain extent of horizontal area, they will be found fractured.

There must have been inequalities in the earth's surface from its earliest consolidation. The radiation of heat, and the necessity of the exterior solid crust conforming to the fluid surface beneath, could not have done otherwise than produce them. Hence M. Élie de Beaumont's theory of the elevation of mountain chains, which rests on the necessity of the earth's crust continually diminishing its capacity, notwithstanding the nearly rigorous constancy of its temperature in order that it should not cease to embrace its internal mass exactly, the temperature of which diminishes sensibly, while the refrigeration of the surface is now nearly insensible*.

Waving for the present the consideration of mountain chains, it will be sufficient to consider that inequalities have existed on the surface of the earth from the earliest times. If we now suppose the solid surface sufficiently cool to support water, that water could not do otherwise than act both chemically and mechanically upon it. The chemical action would be increased by any heat which might still remain in the supporting solid substances, and evaporation would be extensively carried on; so that the more elevated parts being, in all probability, the coolest, and therefore the condensation of aqueous vapour readily taking place upon them, running waters would be established,

* Geological Manual, *Art.* Elevation of Mountains, p. 488.

and a wearing away of the exterior portions the necessary consequence. It might be supposed that the external parts of the cooled crust, composed of various oxidized materials, would be vitreous, and therefore difficult of removal; but we have only to recollect that the vitreous character of igneous products is merely the result of rapid cooling, in order to see that, so far from this being the case under the conditions we have supposed, the necessary slow cooling of the crust would produce a large crystallization of the different substances, and, consequently, a surface by no means difficult of decomposition or removal.

It has been much debated whether the inferior stratified rocks, such as gneiss, mica, slate, &c., have been mechanically or chemically produced; that is, whether they have resulted from the deposit of abraded portions of preexisting rocks mechanically suspended in water, or have been chemically derived from an aqueous or igneous fluid in which their elements were disseminated.

Before, however, we proceed in this inquiry, it becomes necessary to investigate what is generally understood by the words 'strata' and 'stratified', and to see how far 'cleavage' may be confounded with them. A 'stratum' may, perhaps, be strictly defined to be a bed of rock, the upper and under surfaces of which are parallel planes, in all cases where the bed is not perpendicular, but when it is so, then the vertical surfaces would be parallel planes.

This definition is found to be far too rigid for practical purposes, and hence rocks have been termed stratified when divided into beds, the upper and under surfaces of which are not strictly parallel planes, as is seen in the following diagram, representing a section of several beds, *a, b, c, d* and *e,* which would be usually termed strata.

Fig. 1.

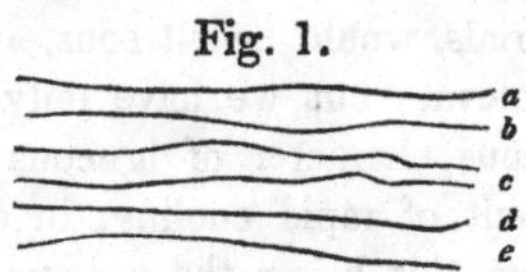

Now irregularity of stratification shows irregularity of formation, and hence becomes an element in inquiries respecting the original production of any given stratified rock. Strata having their upper and under surfaces nearly parallel would seem to require extraordinary tranquillity for their deposition, assuming, for the sake of the argument, that all strata have been deposited. This would be equally true whether they were chemically or mechanically produced. The contrary would hold good with every irregular stratification which may be inferred to result from variations or disturbances in the chemical or mechanical action that produced the rocks thus characterized.

Chemical deposits would take place from solutions saturated with any given substance, from changes among the substances in solution, or from new matter introduced into them, which should produce insoluble compounds. The moment a substance becomes insoluble, it is mechanically suspended in the fluid containing it, and hence will fall more or less rapidly to the bottom, according to its relative specific gravity and volume. All chemical changes in a fluid do not, however, necessarily produce horizontal deposits from it; for, as is well known, pipes conveying mineral waters from place to place are often found incrusted all round. This certainly may be due to the tranquillity arising from the friction of the sides, which will not permit the water to pass so rapidly by them as through the centre of the pipe; but the incrustation on the top is sufficient to show that a chemical deposit may take place against gravity. Numerous crystallizations from saline solutions are produced equally on the sides and bottoms of the vessels containing them, though they may be more abundant on the latter than on the former. Hence chemical deposits may take place on the large scale, and the resulting beds give rise, by the mode in which they repose on other rocks, to the most deceptive appearances.

Fig. 2.

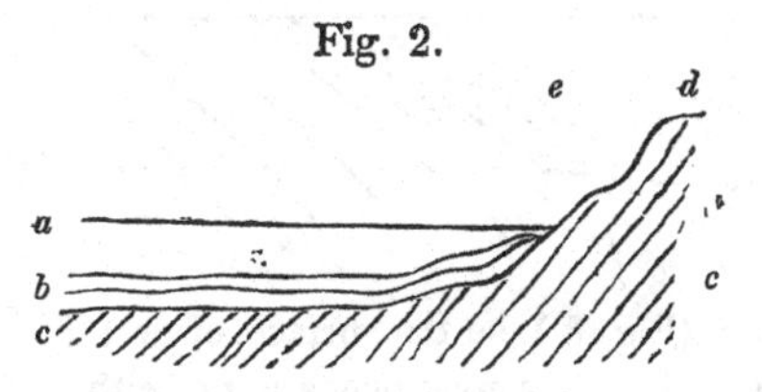

Let *a*, in the annexed figure (Fig. 2.), be the surface of a fluid, such as the sea, from which the beds *b* have been chemically deposited upon the preexisting surface, *c d*, of the stratified rock *c c*. If one of those relative changes of levels so well known to geologists should now take place, by which the beds *b* should be raised above the surface of the sea, there would be some difficulty in deciding whether the upturned appearance of the beds *b* at *e* was due to the protrusion of the mass of *d*, or whether the rocks *b* were originally deposited in the relative position then found. To decide that the tilted character of the beds *b* was due to forces acting from beneath would be evidently wrong in this case. We should therefore exercise some caution in our theoretical decisions when stratified crystalline rocks are thus circumstanced.

A mechanical deposit may also give rise, though in a minor degree, to the same deceptive appearances. If water, containing mechanically suspended matter, move suddenly with a moderate velocity from off shallow into deep water, the resulting stratification will be inclined, even amounting in favourable situations to 40°.

Fig. 3.

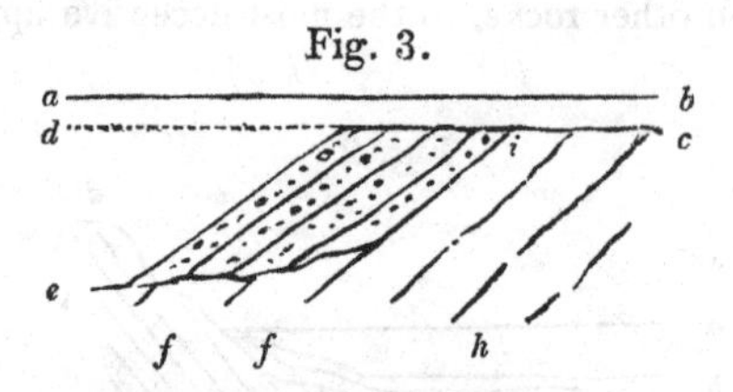

Let *a b* (Fig. 3.) be the surface of water moving over the surface of a rock beneath, *i c*, with sufficient velocity to transport pebbles about one or two inches in diameter, so that they cannot repose on the plane *i c*. It will be clear that they will fall over into the deep water, where they will arrange themselves in strata of high angles, *f f*, being removed from the velocity of the current, which passes on in the direction *c d*. They will fall over into comparatively still water, and therefore arrange themselves according to gravity, and the support that one pebble may afford to the other above it. Mr. Yates and Mr. Lyell have very properly called attention to this mode of stratification*.

Let us suppose, as in the diagram before us, that the rock *h* was stratified in the same manner, (having the same dip and direction as the newly formed beds *ff*,) and that the relative level of water and land should so change that the beds *ff* and the rock *h* rose into dry land, and were exposed, as a section, in a ravine, the surface *e f f* being concealed. It might be inferred that the beds *ff* and the rock *h* had been raised by the same elevating movement, a supposition in this case evidently incorrect, as the beds *ff* have not been raised at all; and if the rock *h* had been tilted up, it was so previous to the existence of the conglomerate beds *ff*.

We have above supposed that pebbles were moved by the current, or stream, over the surface *i c*; let us now consider that sand only is thus transported. If the same velocity of water continued as before, the sand would evidently be shot forward over the edge *i*, and could by no means arrange itself in as highly inclined beds as the pebbles. The finer the sand, the longer it would be finding its way among the particles of still, or comparatively still, water, out of the influence of the superincumbent current. The smaller, therefore, the grain of the sand, the more horizontal would be the resulting strata, and the greater the amount of the fining off at their upper edges. We should thus have, supposing the current to carry forward gravel, sand and finely comminuted

* Yates, "Remarks on the Formation of Alluvial Deposits," Edinburgh New Philosophical Journal, July 1831; and Lyell, Principles of Geology, vol. iii. p. 169. 1833.

matter, a tendency in them all to form beds of various degrees of inclination, the gravel beds rising at the greatest angle, the silt beds being the most horizontal.

If, instead of a stream or current sufficiently powerful to carry forward pebbles, we suppose one merely able gently to move sand over the edge of a high steep bank into deep water, so that the sand should fall grain over grain, the one resting upon the other, and not mechanically suspended in water; we should obtain strata of sand rising at a comparatively high angle.

The natural situations in which circumstances would be analogous to those above noticed are more numerous than might at first sight be supposed. It has been remarked by Mr. Yates*, that strata of this highly inclined character would be produced in certain lakes by the protrusion of detritus where the fal was suddenly into deep water. The same author observes, that similar effects would be produced at the termination of deltas, where they also fall into deep water. This would be the case with that of the Ganges, if the detritus were not too comminuted, and forced forward with such rapidity as to be carried on in the body of the discharged waters, and thus be spread over a considerable area, in beds nearly approaching to horizontality. There are, however, numerous situations off the edges of what are termed soundings, or minor depths of water, on many and extensive lines of coast, where we can imagine the effect of the current but barely sufficient to push

* Yates, "Remarks on the Formation of Alluvial Deposits," Edinburgh New Philosophical Journal, July .1831.

grains of sand over a steep bank, thus causing the formation of strata of sand with an inclination of from 15° to 30°*. Minor effects of this cause are constantly observable on the shifting banks of rivers, and are to be found, as has been often observed, in almost all sandstone rocks. The formation, therefore, of moderately elevated strata, upon a larger scale, is precisely what we should expect under favourable circumstances.

What has been here stated leads to other and important conclusions respecting the formation of mechanical strata; and did we know with certainty the amount of force required to carry forward different kinds of detritus, we might proceed to calculate with some precision the various velocities required to transport, and deposit, given kinds of detrital matter over planes inclined at various angles. Here, however, our data fail us. We have no experiments on which we can implicitly rely respecting the specific gravity and form of detritus which water, with a given velocity, will move on a horizontal surface. It has been stated that a velocity of six inches per second at the bottom will lift fine sand; eight inches, sand as coarse as linseed; and twelve inches, fine gravel; while it requires a velocity of twenty-four inches per second to roll along rounded pebbles an inch in diameter, and three feet per second to sweep along shivery angular stones of the size of an egg†. It will readily strike the reader that the relative specific gravities of

* Mr. Yates has also remarked on the abrupt transition from shallow into deep water off many coasts: "Remarks on the Formation of Alluvial Deposits," Edinburgh New Philosophical Journal, July 1831.

† Encyclopædia Britannica, *Art.* RIVER.

the substances to be transported would greatly modify these effects; but taking the above statement as affording the best approximation that we can at present find, we may proceed to certain general conclusions.

All the velocities enumerated above may be readily found in most rivers, particularly when their beds are moderately inclined, or when they are swept by freshes or sudden rushes of water. Detrital matter, varying in the manner stated, would therefore be also transported, to be here and there deposited, or swept onwards, as the case might be. Circumstances, however, become altered when we regard the extensive deltas of great rivers, large estuaries, or the open sea. Here we find more moderate velocities of moving water, and consequently a diminished transport of all but the finer detritus over considerable distances and areas, at least viewing the subject with reference to the known velocities of water in such situations.

We will not here repeat what we have elsewhere stated* respecting the great improbability that streams of tide, or oceanic currents, can transport any other than fine detrital matter, sand, silt, or mud, except very locally in some narrow channels, or off projecting headlands, where bodies of water are compelled, from the peculiar form and resistances of land, to move at rates greatly exceeding their usual or mean velocities. If exact data fail us in regard to the velocities necessary to transport detrital matter of various kinds by water, they are also sufficiently rare respecting the depths at which ocean currents act.

* Geological Manual, *Art.*, Transporting Power of Tides and Currents, pp. 110, 112.

The only direct experiment of any importance as yet made to ascertain this point is that of Captain Belcher off the west coast of Africa, in latitude 15° 27′ 9″ N., and longitude 17° 31′ 50″ W. He concluded from it that the current moved with nearly the same velocity (0·75 per hour,) at the depth of forty fathoms as on the surface. Now this velocity, supposing it to act upon the bottom, and the foregoing calculations respecting the power of transport correct, would carry forward fine gravel (making due allowance for the differences of statute and nautical miles), though it would be insufficient to roll along rounded pebbles an inch in diameter, which require a velocity of about 1·1 nautical mile per hour.

If we could rely on the transporting powers of water above noticed, we might proceed to calculate the effects of various streams of tide or oceanic currents of depths equal to, or less than, forty fathoms, inferring what might take place at greater depths. But, even assuming that the statement alluded to is a fair approximation to the truth, there is an element of very great importance which must enter into all such calculations : that element is friction. Now the friction of moving water on the bottom over which it flows is very considerable; and no current or stream of tide can have the same velocity on the surface and on any bottom it may pass over. If such were not the case, the tidal waters round most coasts would be a mass of turbid waters. In point of fact, the retardation of flowing waters by friction is so great, that streams of tide with a surface velocity of two miles per hour, and therefore sufficient to transport (according to the calculations above quoted,) angular shivery stones of the size of an egg, pass over sand, and even mud banks, at the depth of a few fathoms, without removing them*.

It would be exceedingly desirable to have more direct evidence, from careful observations on the passage of tidal waters over mud, sand and other banks, of the amount of friction which prevents their removal. Those accustomed to navigate among, or otherwise observe, sandbanks at the mouths of estuaries, must often have observed how, when a falling tide throws the body of ebbing water into numerous channels, forming banks, the edges of such banks begin to suffer removal by the action of the streams upon them. Yet the cutting effect of friction is still very inconsiderable, less than we might expect it to be; and attentive examination will often show that a ripple or surface motion of the water is as destructive as the wearing action of the moving mass of water, which is readily observed to be less near the bank than at a short distance from it. In such situations,

* It is not a little curious to observe how many fish, trout, for instance, avail themselves of the effect of friction on the sides and bottoms of rivers. They usually prefer a certain velocity of water, and therefore employ a given amount of muscular exertion to resist it; but when, from rains, the volume and velocity of the rivers are increased, the fish either sink deeper in the stream or approach the sides, where they regain the former velocity, one which requires no greater amount of muscular exertion than they habitually employ. Without this retardation of water by friction, the fish would be swept out of rivers by freshes, as actually took place from the effects of a hurricane, and therefore an extraordinary cause, in Jamaica. The fall of rain was so great in the hurricane of 1815, that it swept all the fish out of the Yallahs river, one which descends rapidly from the high land of the Blue, St. Andrews, and Port Royal Mountains; and it was considered, ten years afterwards, that there were no fresh-water fish in that river.

also, the different velocities of water, caused by friction on the bottom, may be easily seen before the tide falls sufficiently to leave the banks dry. Still or slow moving water will be apparent over the shoal places, shortly to become dry, while the body of the tide, not being able to move freely over so large a surface as before, runs with greater velocity (unless the time be near low or slack water,) over those parts of the area which are deep, and are in fact the proper channels of the river. Every boatman knows that if he keeps close to a coast or bank when the tide is adverse to his course, he makes more way than when he keeps off, making every difference, in cases where the tides run strong, whether he does or does not reach his destination during that particular ebb or flow of tide, as the case may be.

Now it is important to bear these facts in mind, as they show that when streams of tide pass over sandy bottoms, even with a velocity of two or three miles per hour within a few feet of such bottoms, the moving water does not carry away the sand to any great extent, because there are interposed strata or beds of water which move with a velocity in an inverse ratio to the proximity of the bottom. This retardation is caused by friction; and without it the numerous marine creatures inhabiting mud, sand and shingle bottoms and banks would be unable to exist; they and their retreats would be swept away and transported from place to place by every tide. Thus in this instance, as in every other in nature, design is apparent. Various marine creatures live in mud and sand; the destruction of rocks takes place, and the necessary detritus thus produced is distributed at various depths over the bottom of the sea; but if streams of tide or currents could readily cut up the mud and sand so deposited, this accumulation of detrital matter for the abodes of such marine animals would be useless; and therefore provision is made that it should be comparatively stable.

We must now notice a very important element in all calculations respecting both the production and transport of detritus. This element is the action of waves, known, when they dash on shore, by the name of breakers. This action has often, from the want of proper attention, been confounded with that of streams of tide and currents, in situations where the velocities of the latter were utterly unable to produce the effects observed. Hence many erroneous impressions respecting the destructive and transporting powers of tides and currents. When land rises above the level of the sea, the action of breakers tends constantly to destroy and remove it: when land is low and only rises to the same level, the same action tends constantly to throw detrital matter upon it. There is scarcely any considerable line of coast which does not afford evidence of both these facts. An inexperienced observer could not doubt the destructive action on coasts; but he might be inclined to doubt the piling up of detritus by the sea, thus affording protection against its own ravages, when he sees high banks and sand-hills rising above the sea-level. A little attention would, however, soon show him that the fact is as here stated.

Waves are generally considered as mere undulations of water, the particles of which rise and fall nearly in the same places, communicating movement to other particles, and thus producing vibrations. This view of the subject may be correct enough as regards those waves which are produced by disturbing causes in the body of the water itself, or by throwing a stone into water, resembling the vibrations of the air in the propagation of sound in it. The waves caused by the winds on the ocean must be regarded in another light. They are produced by the friction of the atmosphere on the surface of the sea. No doubt the mere vibratory motion is still in great force, but the superficial particles of water, in contact with the wind, are necessarily forced forward by friction, and surface motion is produced, the velocity of which depends upon the strength or power of the wind on the one hand, and the resistance of the particles of surface water on the other. Every one knows that currents are produced by prevailing winds; the monsoons in the Indian and Chinese seas afford the most complete evidence, if any were wanting, of this fact. Now this could not happen if the friction of the wind on the surface of the sea did not cause a forward motion of the waters of that surface in the direction of the wind.

Those who have experienced heavy gales on the ocean know full well that the water on the summit of waves, thrown forward by the action of the wind, has tremendous forward force; a forward force extending downwards in the wave in proportion to the velocity of the wind above. Hence those fatal accidents to vessels on the open ocean when a heavy wave, or sea as it is termed, strikes a ship, and sweeps everything before it. We should carefully distinguish between such waves and those undulations of water, however large, technically termed swells, which are merely the vibratory portions of waves that always travel with far greater velocity than the particles of water actually propelled forward. These extend far beyond the areas traversed by the winds which have caused the waves, and nothing is more common on the open ocean than the appearance of a swell from a quarter whence there has not been any wind.

The swell is merely an undulation of the water that travels beyond the areas swept by the winds which produce the waves, or that remains in those areas after the wind has ceased to act upon them. These vibrations do not in themselves cause any forward movement of the water. To them such theories as those of Bremontier, Emy and others, which suppose either a succession of vertical or elliptical movements of the particles of water, may be applicable. But the surface water in waves produced by strong winds is clearly propelled forward, the wave breaking from the resistance of the water beneath in the open sea upon the same principle which it does on coasts, with this only difference in the effects, that on coasts, the resistance being solid, the broken wave rushes forward further than it can do when it breaks on a fluid resistance, as it does in the open sea. While, therefore, powerful winds act on waves, the latter are compounded of undulations, or vibrations, and of surface water forced forward by the friction of the moving mass of air above them, while when there are no winds causing friction, waves are mere vibrations. The heavy surf and rolling coast swells, more particularly observable on the shores of islands in large oceans, such as at the Isle of Ascension, St. Helena, the Cape de Verd Islands, &c., are evidently vibrations caused by the continued action of various winds on the surface of a large body of water, and propagated even through water which to the eye appears tranquil. It is not until the proximity of land or shoal water offers resistance that the vibration becomes checked, and is first indicated by a rolling swell, which, rushing on the shore, breaks, and thus that vibration is terminated, as far as the particular coast line extends. Of course these vibrations would be brought to rest, like those producing sound in the air, by a sufficient distance from the disturbing cause; but the case of an open ocean vexed and harassed by a multitude of winds in various parts of its area, is like the air on the surface of country kept in constant agitation by a multitude of sounds on such surface, so that when one shall cease in one situation, another shall commence in another part of the area.

Both waves and swells necessarily disturb the particles of water beneath, to greater or less depths according to their magnitude. The approximative depth to which this disturbance extends is by no means clear: those who have treated on the subject vary much in their estimate. M. Emy contends that it extends to the depth of 500 feet on the Banks of Newfoundland*. It is no easy matter to obtain the necessary data on this head, for the apparent disturbance of a bottom will depend upon the kind of matter of which it may be composed, whether mud, sand, gravel or solid rock. Moreover, when such disturbance takes place, it may not be known to those above, unless they endeavour to obtain, by proper instru-

* Emy, Mouvement des Ondes, Paris 1831, p. 11.

ments, water at different depths, and see when it becomes charged with detrital matter; for the agitation of the water may not be so great as to cause the disturbed bottom of sand or mud to rise high in the sea. Be this as it may, water is evidently discoloured by the action of considerable waves on bottoms of sand and mud, round most coasts, at the depth of about 15 fathoms, particularly when a heavy gale prevails for some hours: thus giving time for the detrital matter to be shaken, as it were, by continued agitation up to the surface.

The mud and sand thus disturbed, and mechanically suspended in water, would return again to their places, or nearly so, when a swell subsided in moderately deep water, though it would be otherwise nearer coasts, where from the great resistance of the bottom there is a forward motion of the water and a breaker the result. Then the mud, sand or gravel would be propelled onwards, necessarily towards the land. The effect of waves driven forward by the action of the wind would be more complicated: the transporting power of the surface would be in the direction of the wind producing the waves. Now the amount of detritus that could be thus forced forward, supposing it to be mechanically suspended up to the surface, would depend upon the depth to which the onward moving water extended, and this again would depend upon the strength and prevalence of the winds producing the waves. As prevalent winds produce currents, our next inquiry is as to the depths to which such currents are felt. Here again exact data fail us. The experiment of Capt. Belcher off the coast of Africa, previously noticed (p. 55), shows us that a current was not much impaired in velocity at the depth of 40 fathoms, and certain circumstances in other situations would lead us to expect that they may extend to at least 80 fathoms. If, therefore, from the action of waves, a bottom be disturbed, the detrital matter thus mechanically suspended in water may be moved onwards in the direction of a current, produced also by the same action. The amount of mechanically suspended matter committed to the moving mass of water would depend upon the shallowness of the sea and the strength of the prevalent wind on the one hand, while the distance to which it would be carried would be in proportion to the depth and velocity of the current on the other.

We must here consider an observation by Mr. Babbage, which though not applicable to the part of the world selected by him for illustration, is yet true in principle. He observes, that if mud, mechanically suspended in water, sink through one foot of that water in an hour, it will be carried by a current, moving at about the rate of three miles an hour, a distance of 1500 miles before it has sunk to the depth of 500 feet*. Now although the velocity here given to a current is not likely to be otherwise than local, still the general principle must be borne in mind. It would be curious to calculate, when from ordinary friction a sand or mud bottom cannot be moved by a current, how far it could be disturbed by the action of waves, and the mud and sand be carried forward in the one case, when it could not in another.

As winds on shore produce waves which gradually

* Babbage, Economy of Manufactures, 2nd edit., p. 51.

increase in the intensity of their transporting powers in proportion to their approach to the land, and as off-shore winds cause calm water close to land, and do not produce waves of importance until a certain distance from it, the tendency of the greater proportion of motion caused by waves round coasts is to throw detritus on shore. The long lines of shingle beaches, and of dunes or sand-hills in front of low-lands, are ample evidences of this fact *.

We now come to the destructive action of waves on land. This constitutes the greatest general force of all those employed by nature for the degradation of that land which rises above the level of the sea. The force of a breaker is little known to those who have not experienced or watched its effects. Its crushing power is in heavy gales so great, that large vessels, unfortunately wrecked, are speedily reduced to fragments, and small vessels are sometimes broken in two by a single blow of a breaker. We are not, therefore, surprised to find that large blocks of rock are easily washed about by breakers, as was well shown at Plymouth during the severe gales of 1824 and 1829, where masses of limestone from 2 to 5 tons in weight were rolled about on the Breakwater like pebbles, and a piece of masonry weighing 7 tons was washed back 10 feet, though it formed a part of the pier in Bovey Sand Bay, and stood 16 feet above

* It was considered useless to repeat here all the circumstances attending the formation of shingle beaches, sandy beaches and dunes, which are noticed under those heads in the Geological Manual, and therefore many interesting facts are here omitted. Those which are, however, thus omitted, do not much affect the question of various kinds of stratification to which the remarks in the text ultimately lead.

an 18 feet spring tide*. These, no doubt, are the effects of extraordinary agitation of the water; but the more general effects of breakers are still, collectively viewed, very considerable. They scoop out by their continual action all those parts of coasts which are soft, if the latter only rise a few feet above the sea-level. Sand, slightly aggregated sandstones, and clays, soon disappear when brought by surface changes into their destructive influence.

Fig. 4.

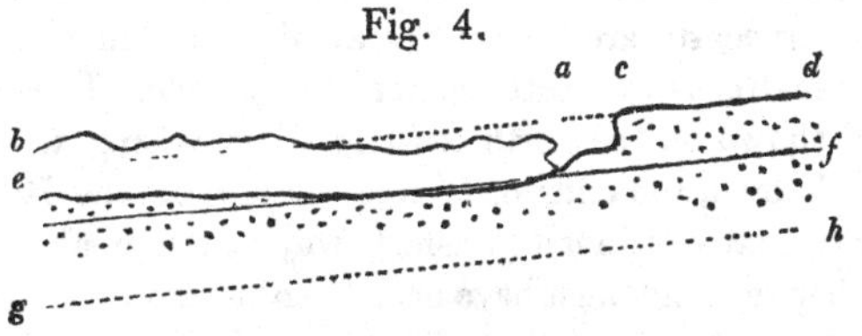

Let *b f*, in the above diagram, represent the level of the sea; *a*, a breaker falling on the shore, and cutting away soft sandstone or clay strata at the cliff *c*. These easily disintegrated rocks would evidently be speedily cut back in the direction *c d*, and the solid matter composing them would be removed back to the sea, to be transported by the power of the tides or currents, as the case may be. Let us suppose that, previous to the existence of this state of things, the surface of the rock now in the act of removal formed a continuous line *e d*, and that this line of surface has been elevated from above the line *g h*, by forces acting from beneath, and which have raised the mass of rocks bodily upwards. While the surface of the rock was at *g h*, it was out of the destructive action of the breakers,

* Geological Manual, p. 82.

but it might have been situated within their piling influence, that is, supposing the form and other circumstances connected with the land in the prolongation of the bed to have been favourable. The moment, however, the elevation took place, so that the line of rocky surface, *e d*, should cross the sea-level, *b f*, the destructive action of the breakers would commence, and the removal of the soft sandstone or clay would be the result.

When clay or other easily disintegrated rocks, forming coasts, are thus cut away, the depth to which such matter is excavated is not considerable. This is well shown on the north coast of Kent, where, as is well known, the cliffs of London clay, near the Reculvers, are continually washed away by the breakers, so that the land must have once been far more extensive than at present, the sea having gained ground at its expense. Historical documents show this to have been matter of fact. Now it appears that the sandbanks in face of the coast are not covered to any great depth by sand. Beneath a few feet clay is found, most probably a continuation of the London clay of the sea-cliffs. So that, if we consider this clay to have been cut down by the action of the breakers, it would appear to be now to a certain degree protected from further excavation.

CHAPTER IV.

Rivers more or less charged with detritus, derived by various causes from the land *, flow out into the sea, particularly during freshes. It has been found that water, even thus circumstanced, is, in almost all cases where attention has been given to the subject, specifically lighter than the sea-water into which these rivers deliver themselves. The relative specific gravity necessarily depends on the relative amount of matter mechanically suspended in the water of the river. Few good experiments have been made to ascertain this point; indeed, it is one which requires considerable attention. It is not sufficient to ascertain the mean annual amount of water passing down a river, and by taking a portion from a particular part of it, see the per-centage of solid matter mechanically suspended in it; it must also be recollected that on the velocity depends the transporting power of the water. In those places, therefore, where the velocity is greatest, the transporting power is greatest. The greatest velocity is necessarily on the surface; and supposing the bed of the river an uniform curve, rising from a central depth to the banks on each side, it would be at equal distances from the banks on either side. Now detritus which could only be carried forward by the central stream, would be stopped in its passage by the next curved stratum of water, if I may so express

* Geological Manual, *Art.* Degradation of Land.

myself. It would, beyond the mere impetus it received when it entered into this stratum, tend to come to rest at the bottom. The resistance of the particles of water in this stratum it would take a certain amount of time to overcome, during which the stratum of water itself would have passed onwards, so that the detritus will gravitate through it in a kind of parabolic curve, which might be calculated, if the depth, relative velocities of the strata of water, and kind of detritus mechanically suspended, were known.

Fig. 5.

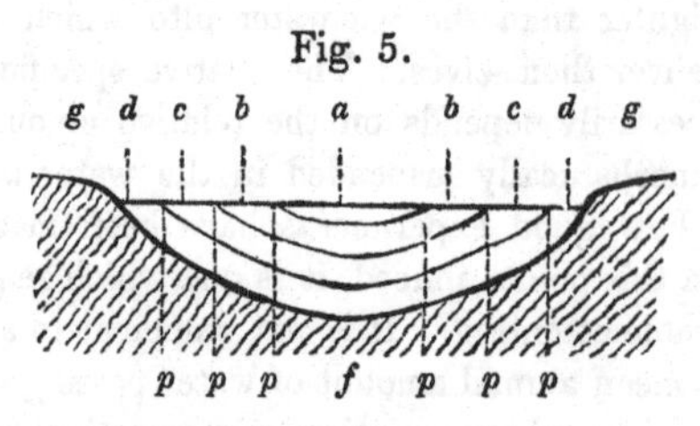

Let the above diagram represent a section of a river, *d, c, b, a, b, c, d,* with its banks *g, g,* the bottom of the bed being at *f*. *a* will be the central stream of water, having the greatest velocity and consequent transporting power. The stratum of water *b, b,* would have less velocity, and therefore less power of transport. The stratum *c, c,* would have still less; and it would be least in the stratum *d, d,* touching the sides and bottom of the river-bed *g, f, g,* where the friction, causing retardation, is greatest. In the above figure the strata of water are, for the purposes of illustration, represented as thick, whereas we must in reality consider them as exceedingly thin.

Fig. 6.

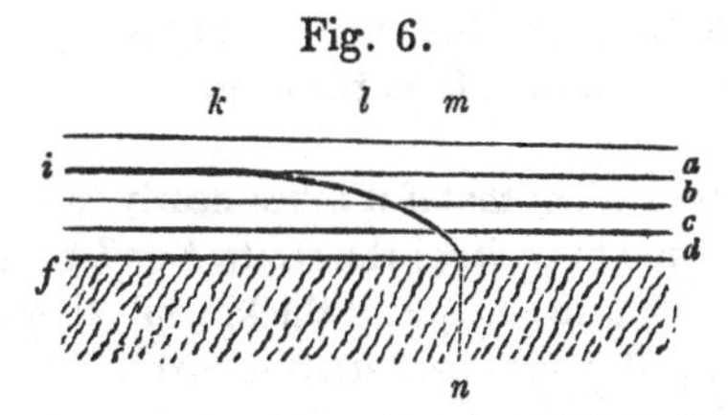

Let *a*, *b*, *c*, *d*, (fig. 6.,) represent a longitudinal section of strata of water, corresponding with the same strata in the cross section, fig. 5. Assuming that the motion among the particles of water in the central stream *a*, is sufficient to keep some given detritus mechanically suspended, precisely as sediment is kept suspended in a vessel containing water by a certain amount of the agitation of such water, it would flow on, sinking but slightly from the action of gravity. Let us now suppose that, either from the action of gravity, a surcharge of mechanically suspended matter, or the friction of the beds or strata of water upon each other causing retardation, a portion of the detritus enters the stratum *b*, where the velocity is insufficient to keep it mechanically suspended; gravity now acts with great force, and the particles of detritus tend to settle perpendicularly downwards. The particles of water offer such resistance in the stratum *b*, that the latter has passed from *k* to *l* (fig. 6.) before the detritus can traverse *b*, and enter the stratum *c*. The velocity of *c* being, however, less than that of *b*, the detrital matter, in traversing through *c*, only passes horizontally from *l* to *m*. From the like causes, in traversing the stratum *d*, the detritus would only pass from *m* to *n* before it came to rest on the bottom *f*.

We should thus have the curved line *i n* (fig. 6.) for the fall of the detritus from the central stream to the bottom.

It will be evident that if it takes nearly equal times for the detritus to traverse the strata *b* and *c* (fig. 6.), supposing them to be equally thick, while they transport it unequally, the lower stratum carrying it a short comparative distance, there will be an accumulation of detritus in the lower strata; consequently, the nearer the bottom, the larger would be the amount of the detritus we should expect to find. Experience teaches us that this is also the fact. It may now be said, that if we take all these circumstances into consideration, and strike a kind of mean, we shall have no difficulty in estimating the amount of detritus carried down by a river at any given time. This would be the case if we had only to take into account a particular perpendicular section of water of small horizontal thickness, such as we may consider fig. 6. to be. If, however, we now consider the river to be divided into numerous sections of the same kind, each descending perpendicularly to the bottom, as is represented in the cross section, fig. 5., by the lines *p, p, p, p, p, p,* we shall have two series, one on each side of the central section, the terms of which shall not agree either in respect to the velocities of the water, the power of transport, or in the amount of detritus contained in them. That the velocities would be unequal, is readily seen by a glance at fig. 5., for the perpendicular sections unequally cut through different strata of water. The transporting powers are consequently different. The less amount of detritus would not wholly depend on this cause, for the stratum *d* is uncovered by the others near the banks *g*, *g*; hence it has not received any detritus in that portion of it, from the fall of mechanically suspended matter through *c* or *b*.

It will have been seen that, so far from being easy to determine, a calculation respecting the amount of detrital matter carried down, even at a given time, by a river, whose section should be as uniform as that represented in fig. 5, would be a work of extreme difficulty. Now a section so simple can scarcely ever be obtained in nature; on the contrary, a form of bed so favourable for calculation must be considered as one of the greatest rarity. The usual sections of river-beds are far more complicated. When, therefore, we regard what has been above stated respecting the difficulty that would be encountered, even in estimating the amount of transported matter carried onwards by a river at any given time, and reflect that such amount is constantly varying, we shall perceive that to obtain anything like a tolerable annual mean, requires more experiment, time, and calculation, than can be afforded by any than those who are favourably situated for the purpose, and who are willing to devote themselves to it.

The usual method of estimating the general amount of detritus borne down by any great river which may form a delta, is by calculating the increased superficial area of such delta during any given time, such as one year, or one hundred years. The superficial increase of a delta by no means affords the necessary information, unless the depth of water into which the delta may protrude be also taken into account. If transported matter, carried down by the river, be added to the external fall of a delta, the amount of such matter would be the same, all other circumstances being similar, whether we consider the length of the superficial increase as 2 and the depth as 4, or the former as 1 and the latter as 8. It follows that a delta which shall increase a certain length superficially in a given time in a given depth of water, will have received, all other circumstances being the same, as great an addition of solid matter as another delta protruded into half the same depth of water, the superficial advance of which should be twice as great in the same time.

The distances to which river-water, more or less charged with detritus, would flow over sea-water, will depend upon a variety of obvious circumstances. Captain Sabine found discoloured water, supposed to be that of the Amazons, three hundred miles distant in the ocean from the embouchure of that river. It was about 126 feet deep. Its specific gravity was = 1·0204, and the specific gravity of the sea-water = 1·0262. This appears to be the greatest distance from land at which river-water has been detected on the surface of the ocean. If rivers containing mechanically suspended detritus flowed over sea-water in lines which, in general terms, might be called straight, the deposit of transported matter which they carried out would also be in straight lines. If, however, they be turned aside by an ocean current, as was the case with that observed by Captain Sabine, the detritus would be thrown down, and cover an area corresponding, in a great degree, with the sweep which the river has been compelled to make out of the course that its impulse, when discharged from its embouchure, might lead it to take. Supposing the velocity with which this river-water was moving has been correctly estimated at about three miles per hour, it is not a little curious to consider that the agitation and resistance of its particles should be sufficient to keep finely comminuted solid matter mechanically suspended, so that it would not be disposed freely to part with it, except at its junction with the sea-water over which it flows, and where, from friction, it is sufficiently retarded. So that a river, if it can preserve a given amount of velocity flowing over the sea, may deposit no very large amount of mechanically suspended detritus in its course from the embouchure to the spot where it is ultimately stopped. Still, however, though the deposit may not be so abundant as at first sight would appear probable, the constant accumulation of matter, however inconsiderable at any given time, must produce an appreciable effect during the lapse of ages.

Though mechanically suspended detritus is thus carried far out into the sea in certain favourable situations, the greater amount will always be accumulated near the embouchures of rivers which transport it. The river is most loaded with foreign matter near its bottom, where also the retardation from friction is greatest; therefore we should anticipate that this effect must be produced.

We have now considered the various means by which detritus is carried into the sea, and, to a certain extent, the manner in which it is moved by currents or streams of tide. It now becomes necessary to notice the specific gravities of various substances most commonly transported.

Specific gravities of Minerals most commonly entering into the composition of Rocks.*

N. Selenite.........2·2 to 2·4	B. Steatite .... 2·63
D. B. Calcareous spar (Iceland)...... 2·71	H. Chlorite .... 2·71
H. Arragonite .... 2·93	Hy. Chiastolite........ 2·94
D. B. Rock Crystal... 2·65	H. Hypersthene ...... 3·38
D.B. Quartz (common) 2·63	D. B. Hornblende ... 3·27
B. Flint (chalk) ...... 2·59	D. B. Augite (black). 3·16
D. B. Chert .... 2·64	D. B. Olivine (grains) 3·39
H. Felspar......2·53 to 2·60	H. Leucite .... 2·48
H. Albite ...... 2·61 to 2·68	H. Garnet...... 3·61 to 4·20
D. B. Felspar (Labrador) .... 2·56	H. Shorl, or Tourmaline .... 3·07
H. Mica .... 2·94	H. Diallage .... 3·25
D. B. Talc .... 2·76	N. Iron pyrites 4·5 to 4·90

Specific gravities of Rocks.

D. B.	Calcaire grossier (Paris)	2·62
D. B.	Chalk, pure (Sussex)	2·49
D. B.	Upper green sand (Wilts)	2·47
D. B.	Lower green sand (Wilts)	2·61
D. B.	Portland oolite (Portland)	2·55
D. B.	Forest marble (Pickwall)	2·72
D. B.	Bath oolite (Bath)	2·47
D. B.	Stonesfield slate (near Stow-in-the-Wold)	2·66
D. B.	Lias limestone (Lyme Regis)	2·64
D. B.	Red marl of the new red sandstone (Devon)	2·61
D. B.	Muschelkalk, fossiliferous (Göttingen)	2·62
D. B.	Coal sandstone, Pennant (Bristol)	2·60
D. B.	Coal shale, with impressions of ferns (Newcastle)	2·59
	Bituminous coal (Newcastle)	1·27
D. B.	Millstone grit (Bristol)	2·58
D. B.	Carboniferous limestone (Bristol)	2·75
R.	Carboniferous limestone (Belgium)	2·72

* The letters prefixed to the various substances in these tables show by whom the respective specific gravities have been ascertained. Hy., Haüy; H., Haidinger; B., Breithaupt; R., Rondelet; B., Brisson; N., Naumann. Those marked D. B. were taken by myself.

D. B.	Old red sandstone, micaceous (Herefordshire)	2·69
D. B.	Old red sandstone (Worcestershire)	2·65
D. B.	Grauwacke (Hartz)	2·64
D. B.	Grauwacke, common (Ilfracombe, Devon)	2·69
D. B.	Grauwacke, calcareous (Ilfracombe)	2·77
D. B.	Grauwacke (Snowdon)	2·76
D. B.	Argillaceous slate, grauwacke (Devon)	2·81
D. B.	Carrara marble, pure	2·70
D. B.	Mica slate (Scotland)	2·69
D. B.	Gneiss (Freyberg)	2·72
D. B.	Domite (Puys de Dôme)	2·37
D. B.	Trachyte (Auvergne)	2·42
D. B.	Basalt (Scotland)	2·78
R.	Basalt (Auvergne)	2·88
D. B.	Basalt (Giant's Causeway)	2·91
D. B.	Greenstones, various (different countries)	2·69 to 2·95
D. B.	Sienite (Dresden)	2·74
D. B.	Porphyry (Saxony)	2·62
D. B.	Serpentine (Lizard, Cornwall)	2·58
D. B.	Diallage rock (Lizard, Cornwall)	3·03
D. B.	Hypersthene rock (Cocks' Tor, Dartmoor)	2·88
R.	Granite, green * (Vosges)	2·85
R.	Granite, grey (Brittany)	2·74
R.	Granite (Normandy)	2·66
D. B.	Granite, mica scarce (Scotland)	2·62
D. B.	Granite (Heytor, Devon)	2·66

Specific gravities of some Shells†.

Land.

Helix Pomatia	2·82	Auricula bovina	2·84
Bulinus decollatus	2·85	Helix citrina	2·87
——— undatus	2·85		

Fresh-water.

Unio cardisce	2·79	Paludina	2·82
—— cicatricosus	2·80	Cyrena Sumatrensis	2·82

* Query greenstone?

† The specific gravities in this table were ascertained by myself.

Marine.

Argonauta tuberculosus	2·43	Strombus Gigas	2·77
Nautilus umbilicatus	2·64	Chiton	2·79
Ianthina communis	2·66	Pholas crispata	2·82
Lithodomus Dactylus	2·67	Cytherea maculata	2·83
Teredo (great, East Indies)	2·68	Bulla	2·83
Haliotis tuberculatus	2·70	Voluta musica	2·83
Cyprina vulgaris	2·77	Cassis Testiculus	2·83
Mytilus bilocularis	2·77	Strombus Gibberulus	2·83
		Pyrula Melongena	2·84
		Tellina radiata	2·85*

Before we proceed to consider the subject for which the above tables have been introduced, let us for a moment regard the data they afford us for a rough approximation towards the mean density of the earth's mineral crust. This density has been usually estimated at 2·5. It would, however, appear that this estimate is somewhat too low, and that 2·6 would be a nearer approximation. If we take a mean depth of fifty miles beneath the bed of the sea and the

* It can scarcely escape the observation of the reader, that, while the specific gravities of the land shells enumerated is generally greatest, the densities of the *floating* marine shells is much the smallest. The design of the difference is obvious. The land shells have to contend with all changes of climate, and to resist the action of the atmosphere, while at the same time they are thin for the purpose of easy transport: their density is therefore greatest. The Argonaut, Nautilus, and creatures of the like habits, require as light shells as may be consistent with the requisite strength; the relative specific gravity of such shells is consequently small. The greatest observed density was that of a *Helix*, the smallest that of an *Argonaut*. The shell of the Ianthina, a floating molluscous creature, is among the smallest densities. The specific gravity of all the land-shells examined was greater than that of Carrara marble; in general more approaching that of Arragonite. The fresh-water and marine shells, with the exception of the Argonaut, Nautilus, Ianthina, Lithodomus, Haliotis, and great radiated crystalline Teredo from the East Indies, exceeded Carrara marble in density. This marble and the Haliotis are of equal specific gravities.

surface of dry land, we might expect that gneiss, mica slate, hornblende rocks, granites, greenstones, and rocks of that character, would constitute about ·8 of the whole mineral mass. Now all these rocks exceed the specific gravity of 2·6. If we estimate granite, gneiss and mica slate as equal to ·5 of the mass, such portion would have the mean density of about 2·7. Taking greenstone and trappean rocks at about ·1, and including diallage rock and serpentine among them, we should obtain about 2·76 for their mean specific gravity. We cannot estimate the mean density of the other rocks of this portion, such as hornblende rocks, chlorite slates, quartz rocks, clay slates, &c., at less than 2·65; indeed a more detailed calculation would probably make it greater; but taking it at 2·65, we should obtain 2·695 as the mean density of ·8 of the earth's mineral crust. Of the remaining ·2, the density would be very variable. Certain calcareous rocks, such as chalk and the common oolites, would be comparatively light, as would also be the case with some sandstones, such as some in the cretaceous series. Some lavas thrown out in the atmosphere, or beneath moderate depths of water, would also have comparatively small specific gravity: the same would also be the case with domites, trachytes, and rocks of that class. Imbedded vegetables forming lignite and coal, must also be included in this account. If we take these rocks as forming ·1 of the mineral mass under consideration, and that its mean is about 2·45, we shall probably not commit any gross error in this rough calculation. The remaining ·1 will be at least equal to 2·67. The mean density of the old red sandstone and grauwacke would, from the preceding tables, be 2·71. Now these rocks, or others quite analogous to them in mineralogical structure, though not always of the same age, such, for instance, as many in the Alps, Oural, Himalaya, and other great ranges of mountains, would constitute a large proportion of the ·1 under consideration. The mean density of the various compact limestones cannot be much less than 2·66: when fossiliferous or filled with shells, they will not be lighter; on the contrary, their densities will in most cases be then increased, as will be readily understood by reference to the specific gravities of shells. To this portion we must add the basalts and rocks of that class, often, as in India, occupying large tracts of country, and common in many volcanic districts. We cannot estimate the mean density of the basalts at less than 2·8. Taking the grauwacke and rocks of that structure at ·4 of the ·1 of the earth's crust under discussion, the compact limestones at ·2, the compact sandstones at ·3, (to which we may assign a mean density of 2·6,) and the basaltic rocks at ·1, we obtain a mean of 2·67 for this portion. We thus get about 2·59 for the mean density of the mineral crust of the globe. The increased density from the pressure of the superincumbent mineral masses at various depths is here neglected; but when we regard this, and consider that in this rough calculation the proportions of the more dense rocks have been rather under- than over-estimated, we may probably take 2·6 as a fair approximation to the truth*.

To return, however, from this digression: it will

* It is my intention more fully to investigate the subject elsewhere, when, from more abundant data and more minute detail than could be admitted into this volume, it is hoped that a more certain as well as a closer approximation may be attained.

be obvious that equal volumes of the various substances enumerated in the foregoing tables will not be transported equal distances by a given velocity of water, all other things being equal. If we suppose equal volumes and like forms of mica, quartz, felspar and hornblende to be mechanically suspended in moving water, and gradually deposited from it, the hornblende would first come to rest, then the mica, afterwards the quartz, and finally the felspar. If equal volumes and like forms of shells, such as *Helix citrina, Strombus Gigàs*, and *Argonauta tuberculosus*, were committed to a current of water, in the shape of fine grains, such as we often find in nature, the fragments of *Helix* would be first deposited, then those of the *Strombus*, and afterwards those of the *Argonaut*. The specific gravities of the shells above given are not sufficiently numerous to justify any very general conclusions; but as far as they go, they tend to show that if equal volumes and like forms of land, freshwater and marine shells were mechanically suspended in moving water, and gradually fell to the bottom of the sea or river, as the case might be, the land shells would be deposited before the others.

If we consider that a river may carry equal volumes of land shells, quartz and felspar into the sea, the felspar would be transported the greatest distances, while the shells would be thrown down nearest the land, and the quartz would occupy a mean place between the two: so that, under these conditions, the resulting deposit, all others being equal, would be most calcareous nearest the land, and most siliceous further out, while the furthest zone of all, if we may use the expression, would be most argillaceous, assuming a certain amount of mixture in the substances deposited.

Supposing the matter carried outwards by a river, highly charged with detritus, was composed of selenite, felspar, quartz, calcareous spar, talc, chlorite, mica, hornblende, olivine and garnet, with equal volumes and like forms, it would tend to come to rest in the inverse order of the minerals here enumerated. If, therefore, running waters, such as rivers, flow off a country of common granite, composed of quartz, felspar and mica, the velocities of the rivers being sufficient to carry the detritus into the sea, we should expect to find the micaceous matter predominate near the land, while the comminuted felspar would prevail at the outer edge of the deposited detritus. Now it must have frequently struck those whose attention has been given to the detritus carried down by rivers, that in granitic districts the tendency to arrangement is such as above stated, due allowance being made for variations in the form and volume of the substances transported. The mica endeavours to remain last; but as its volume is generally less than that of the quartz or felspar, it is carried onwards, sometimes with, sometimes beyond the other minerals, as must happen in all cases where volume or form is sufficient to equalize or overpower the effects of unequal specific gravity.

Sandstones are frequently micaceous, and are often rendered schistose by plates of mica. For the most part such sandstones are siliceous. It will have been seen that equal volumes of mica and quartz, mechanically suspended in water, would not descend in equal times to the bottom, all other things being equal. May we not here approximate towards an explanation of those alternations of mica and quartz grains so common in many rocks? The grains of mica are often, in such cases, of less volume than those of quartz; but supposing their volume the same, and that a liquid, such as the sea, be supplied at equal intervals with such detritus by another liquid charged with it, such as a river, there will be a tendency to have alternations of mica and quartz, for the first charge of mixed detritus will have fallen a certain distance before the next is delivered into the sea. We must not, however, forget that form would greatly influence a deposit of this nature. Though mica is specifically heavier than quartz, the tabular form of the former, and the rounded character of the detrital grains of the latter, would not allow these minerals to descend through water equally in equal times, even though their volume might be the same. If various mineral substances be mechanically suspended in the same liquid at the same time, the nature of the deposit from such a mixture would depend upon the relative densities, form and volume of each. Arrived at the bottom, the small fragments of mica would rest upon their flat surfaces, if they even first fell upon their edges, and the result would be the lamellar structure we observe in micaceous sandstones.

The differences in specific gravities must necessarily produce great effect in the deposit of detritus of a larger kind, derived from various rocks. Thus equal volumes of Bath oolite and carboniferous limestone, supposing the forms the same, would not be transported equal distances by equal forces: the Bath oolite would be carried the greatest distance. Equal volumes of trachyte, granite and basalt would, under similar circumstances, be transported unequal distances. If equal volumes of granite from Normandy or Heytor, Devon, and of Stonesfield slate from Stow-in-the-Wold, were carried onwards by moving water, the tendency to come to rest would be equal; the difference in their actually doing so would depend upon their form. Most of the granitic rocks would travel further than many of the compact limestones, (particularly when fossiliferous, such as the forest marble,) certain argillaceous slates, many of the grauwacke rocks, and the mass of basalts, greenstones, hypersthene and diallage rocks.

Though the specific gravities of various minerals and rocks are thus important, and may lead to valuable information when equal volumes of matter are concerned, and we seek to discover the direction whence any mechanical rock has been derived, we must be careful to pay due attention to the differences of form and volume, when such exist, otherwise we may arrive at very erroneous conclusions. We must in the first place recollect, that when the same force acts on different moving bodies, it communicates velocities to them which are in the inverse ratio of their masses, or of the quantity of matter of which they are composed. Therefore, in estimating the forces which may have transported any conglomerate or sandstone beds, we must pay proper attention to the mixture of the mass, and consider how the force has acted. We should see whether it appears to have been sufficient to have driven all the detritus, great and small, on before it,—the smaller parts being entangled in the eddies produced by the larger bodies, which did not move with equal velocities,—or whether the detrital matter be arranged with reference to specific gravities, forms and volumes. When we detect large rounded masses, or pebbles of large size, mixed up pell-mell with detrital matter of much smaller volume (as at *a*, fig. 7.), we may infer that the velocity which

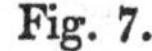
Fig. 7.

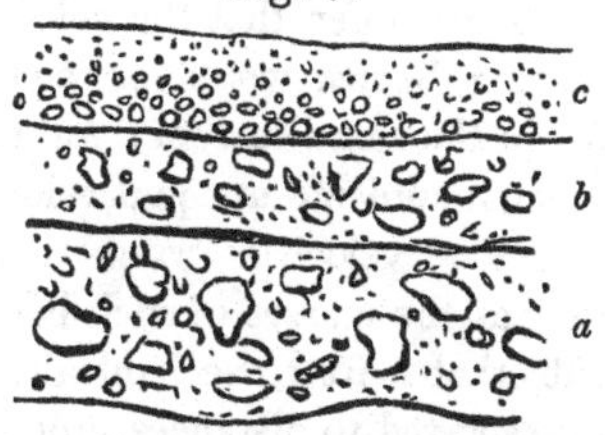

transported the mass was considerable, the larger bodies being, for the time, mechanically suspended in the water. When we find pebbles in such conglomerate rocks arranged with reference to their respective volumes, forms and densities, we have no evidence that the larger bodies have ever been suspended in the water which forced them onwards. They may have been merely rolled forwards on the bottom.

The two kinds of arrangements may sometimes be observed in rivers in which violent floods occur. The deposit from the sudden rush of water generally consists of a mixture of great and small pebbles, sand and silt, much the same as is represented at *a* and *b*, fig. 7.; but when the waters are only sufficient to roll the pebbles over the bottom, there is always a tendency to carry onwards the finer matter, which necessarily becomes exposed to the action of the current by the removal of the pebbles that might otherwise protect it. The finer portions being thus removed, the larger bodies arrange themselves as at the bottom of the bed *c*. No doubt there is the same tendency to separate the finer matter from the larger bodies, when all are, for the time, mechanically suspended in water; but the case is so far different that the larger and smaller bodies are borne onwards together. We do not infer that the velocities of the smaller and larger bodies would be equal, though borne onwards in the same current; for the original impulse would communicate unequal velocities to the bodies moved. The velocities would be, as above stated, inversely as the masses of such bodies. There can be no doubt that matter, mechanically suspended in water, is transported to distances proportionate to its volume, form and density; yet in cases where detrital or other matter is suddenly caught up, as it were, by a great rush of waters, and for a time mechanically suspended in them, it is curious to observe how confusedly mixed the resulting deposit is when thrown down where the rapidity of the current ceases. When we look at the surface of masses of detritus so borne down, we might often be led to suppose that the masses themselves were composed solely of large pebbles, as little else may be visible upon such surfaces; but artificial or natural sections generally exhibit a confused assemblage of stones and sand, arranged without much regard to specific gravities, forms, or volumes. These confused assemblages of transported matter seem always the result of tumultuous action, and the deposit from waters highly charged with rocky matter of most variable dimensions. Although such rock-charged masses of water necessarily move as a whole with great rapidity, their internal motions are very different, from the unequal velocities of the substances transported, and the consequent confusion of eddies and counter-eddies, and the whirling of the substances mechanically suspended; so that when suddenly brought to rest, as frequently happens in such cases, there is a pell-mell mixture of large and small bodies, and of every kind of rock carried onwards. The surface of these deposited masses necessarily becomes washed, and the smaller bodies removed; for, when brought to rest, the water, before it passes off from them, acts with considerable power, precisely as it would on the bottom of a river.

When, therefore, we examine conglomerates, and find that the beds contain fragments of various kinds of rocks, or even of the same rock, arranged without regard to density, form, or volume, we may infer that they have been thrown down from a violent rush and the tumultuous action of water. On the contrary, when we find the pebbles or fragments with something of the order which they would present if deposited in a comparatively quiet manner, we may infer that if they have been mechanically suspended in water, they have not been thrown violently upon the surface on which they rest, but have gradually separated out from the moving water as the velocities of the latter decreased. We may often remark that the fragments are more angular in those beds in which they are confusedly arranged and of very various sizes, than in those where the pebbles or fragments are of more uniform sizes, and deposited in an order more corresponding to their relative densities and volume.

These remarks are intended only with reference to

conglomerates, either distributed over considerable areas, or in situations where we may consider a littoral deposit improbable. Lines of conglomerate, composed of materials of very unequal volumes, may readily be formed on coasts, particularly when steep.

Fig. 8.

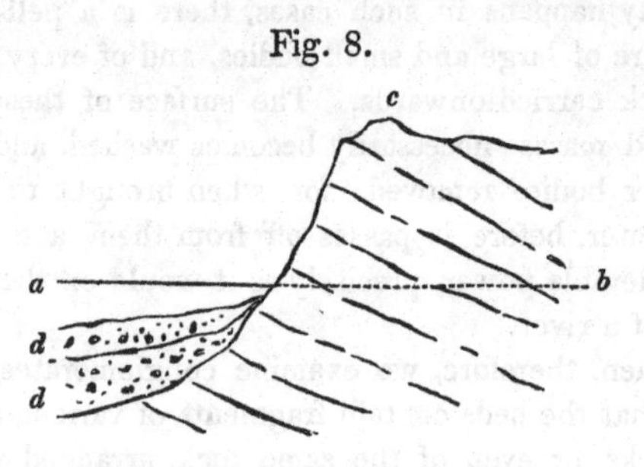

Let *a b* (fig. 8.) be the level of the sea, and *c* a cliff, which, both by decomposition and the action of the sea upon it, produces fragments of unequal size, that fall into the water at its base. The action of the waves will tend to arrange them more horizontally than would happen if they fell upon dry land; and from the unequal action of the waves, and the various size of the fragments, the arrangement can scarcely be otherwise than confused, much resembling that noticed above as resulting from a violent and sudden rush of water, highly charged with variously sized detritus. From the nature of things, however, beds of this kind could not be long continuous in directions at right angles to the lines of coast on which they have been produced. We may hence distinguish them from those distributed over wide areas, and where there is no reason to suppose that a retreat of cliffs, by decomposition, or the action of the sea, could have occasioned them.

As we may infer a large and confused deposit to be the result of violent and tumultuous action of water, so may we consider that a conglomerate or sandstone, in which the volume of the pebbles or grains of sand is equal, or nearly equal, in each bed, is the consequence of a more regular or continuous movement of water, from which mechanically suspended detritus is thrown down. There would necessarily be much uniformity in beds of conglomerate produced by a given velocity of water forcing pebbles before it; for a given velocity would only drive fragments of a certain density, volume and form before it, leaving those behind which were more dense, of greater volume, or of less moveable forms, while all detritus more favourable for transport would be carried onwards. Particles of finer detritus must become mixed with the larger, when the latter, from diminished velocity in the water, come to rest; for around the exposed surfaces of the larger bodies the water will be so retarded, that the smaller will be deposited from it, and introduced among the interstices of the fragments and pebbles, as well as in those situations where small eddies would be formed. The resulting beds would present a uniformity corresponding to the uniform velocity of the moving water. When variations were produced in the latter, corresponding changes would be effected in the former; and we should have beds which, though to a certain extent uniform in themselves, would differ in the magnitude of the constituent detritus, as is represented by the beds *a*, *b* and *c*, fig. 9.

Fig. 9.

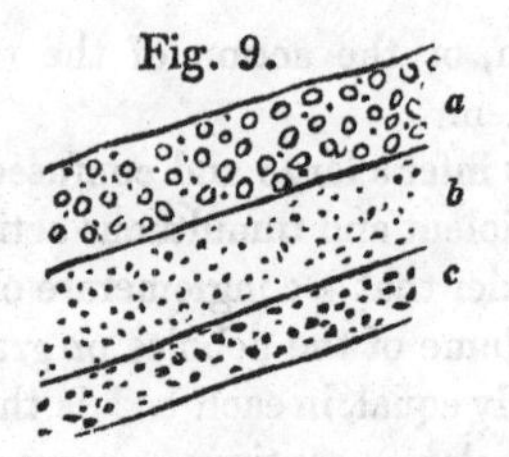

Most rocks, mechanically produced, afford marks of the unequal action of the water from whence they have been deposited. This is particularly observable in sandstones.

The annexed wood-cut (fig. 10.) represents a section of a kind by no means uncommon, and is one

Fig. 10.

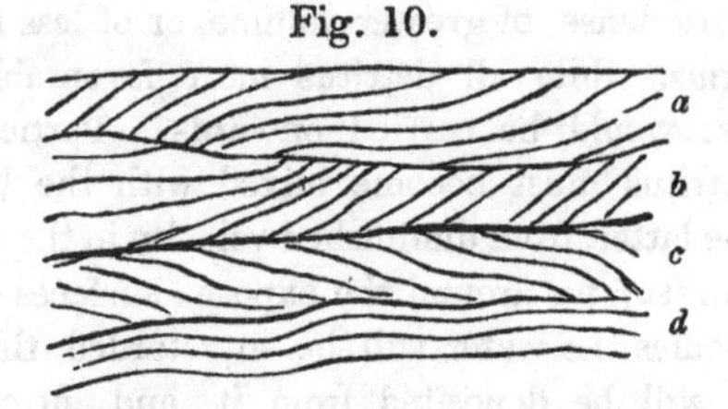

strongly characteristic of the forcing action of water over a bottom, not a quiet deposition of matter from a state of mechanical suspension. We may suppose *d* a sandstone bed resulting from the quiet deposition of grains from water in which they were mechanically suspended: the laminæ may be considered sufficiently horizontal. This, however, could scarcely be the case with *c*, *b* and *a*, where the grains of sand must have been forced over each other upon the bottom, and have not descended solely from the action of gravity until they came to rest, remaining afterwards undisturbed until quietly covered over by other grains which had fallen in the same manner. We must carefully distinguish between the diagonal or waved lines here under consideration, and the cleavage of rocks to be noticed in the sequel; but when we are certain that the lines in question are not those of cleavage, we may conclude that the rock has not been formed by a superior current of water transporting detritus which subsequently fell to the bottom through still water. It must have been caused by the pushing action of moving water on grains of sand after they had touched the bottom. Variations in the transport are marked by changes in the directions and forms of the sectional lines, while large breaks in the deposit are shown by the greater lines, which are sections of planes, that divide the deposit into beds, such as *a*, *b* and *c*, fig. 10.

A small wavy or ripply surface is by no means uncommon in mechanical rocks, and may result from the friction of water on an arenaceous bottom, in the manner commonly observable on sands laid bare at low tides, or on loose sandy districts where the wind performs the same office as the water on the bottom beneath it, as has been noticed by Mr. Lyell*. The effect of waves, particularly those which act with transporting force on the bottom, is to raise lines of ripple marks. The particles at the bottom are then, indeed, forced onwards by friction, the lines of little elevations being modified by the vibratory action also produced in the water at the same time. The ripple marks, modified by vibrations in the water, may be

* Principles of Geology, vol. iii. p. 176.

expected only in moderate depths, to which the vibrations of a wave might extend, while the little elevated lines caused by the simple friction of a mass of water upon yielding sand may be formed at any depths in which water could move with sufficient force.

CHAPTER V.

We can scarcely conceive that detritus, particularly when finely comminuted, and then most favourable for the exertion of chemical affinity, can be in juxtaposition, and kept moist, and even wet, for long periods of time, without suffering some chemical change. The abundance of crystals of selenite and iron pyrites in clays, that have evidently been deposited from water in which finely comminuted matter had been mechanically suspended, clearly show that chemical affinity has there been sufficiently strong to overcome the attraction of cohesion, which, to a certain extent, must have existed among the particles of deposited matter.

Such effects must, of necessity, constantly vary. It would be difficult to find any two deposits from fine detritus, mechanically suspended in water, which should be perfectly the same. Chemical changes in them would therefore be produced under the influence of different circumstances. Rocks must also receive considerable chemical modification by the percolation of water through them. Those which, from geological changes in the relative level of sea and land, are elevated above the sea, would not be placed under the same conditions as those which remained beneath the sea. There is scarcely any spring water, such as may be considered the result of the percolation of atmospheric water through rocks, that does not contain some mineral substances in solution, which it must have procured in its passage through the rocks. Now, though this quantity may be small, when we regard the composition of any particular spring water, yet when we consider the soluble matter contained in the spring waters of any given 1000 square miles of country, and that this subtraction of matter from rocks has been going on for ages, we may readily conceive that the amount of chemical change may be greater than at first sight we might anticipate. We may also infer that the more soluble portions of rocks have a constant tendency to be removed when exposed, not only to direct atmospheric influences, but also to the percolation of rain water through them, so that most rocks would experience great difficulty in resisting chemical changes of this kind, and of preserving their original chemical nature, more particularly when elevated into the atmosphere.

Dr. Turner has shown that in the decomposition of the felspathic rocks, producing porcelain clay, the quantity of silica carried off by solution was enormous. He attributes this loss to the freedom with which it could be dissolved when exposed to the united action of water and alkali at the moment of passing from the state of combination which constitutes felspar. The same author considers "that every 2 equivalents of alumina, present in porcelain clay along with $3\frac{1}{2}$ of silica, corresponded in the original felspar, from which it was derived, to 12 equivalents of silica and 1 of potash*."

The great mass of mechanical rocks has been deposited in the sea. Hence we might generally expect

* "On the Chemistry of Geology," London and Edinburgh Phil. Mag., July 1833.

to find at least the predominant salts of the sea disseminated through such rocks, so as to be readily detected. This, however, is not the case. Now it is difficult to conceive how this saline matter is subtracted from them to such an extent as we find it is. We may suppose that when elevated from beneath the sea into the atmosphere, the rain water, which then has a tendency to percolate through such rocks, would carry off a large proportion of the saline matter in solution. This would undoubtedly be a slow process, but, continued through a long series of ages, it would appear to be certain in its effects. This hypothesis would not, however, account for the absence of marine salts in rocks beneath the level of the sea.

The collective amount of soluble matter carried by rivers into the sea, and derived from rocks by the percolation of water through them, must be very considerable, more particularly when we measure time by centuries, and not by single years. Rivers in which a large amount of finely comminuted detritus is mechanically suspended, cannot fail also to contain matter in solution. The very detritus itself is likely to be acted on by the water in which it is suspended, even if the latter was not impregnated with soluble matter derived from the rocks whence its springs issued. Calcareous detritus would be liable more particularly to be so acted on, as also the fine comminuted particles of felspathic rocks.

All this matter must be carried somewhere. Many changes must be produced by the action of the various substances thus carried out into the ocean, not only on each other, but also on the saline contents of the sea. Chemical deposits cannot fail to be produced, even of kinds which we do not anticipate from our present knowledge either of the relative amount of substances thus situated, or of the possible combinations they may enter into. The chemical deposits which we are most accustomed to study, as it were, in the act of formation, have not been produced under the same circumstances. They are chiefly calcareous and derived from the waters of rivers, freshwater lakes, or thermal springs. The few natural deposits of silica and sulphate of lime which we observe now taking place are chiefly from the latter. Of the chemical rocks which may be now forming in the sea we know almost nothing; yet all the soluble matter conveyed into it from the land, independently of the soluble substances thrown into it from volcanic action beneath its surface, must be productive of some result.

There are also certain aggregations of particles in rocks, mechanically produced, which must have resulted from mutual attractions among each other, and which are not a little remarkable. It not unfrequently happens that in clays, containing disseminated carbonate of lime, there are nodules more calcareous than the other parts, and which we readily perceive are not bodies rounded previous to deposition, although at a distance they have that appearance. We will select for illustration some nodules of this kind in the lias of Lyme Regis, though such bodies are sufficiently common elsewhere.

Fig. 11.

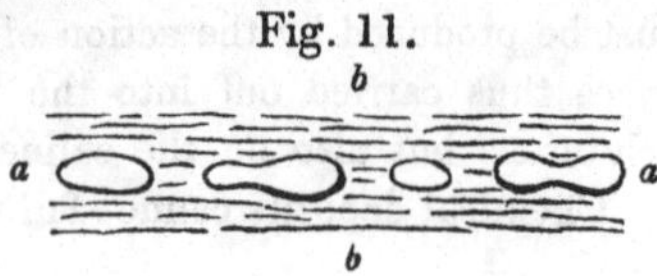

The nodules *a a* (fig. 11.) range in a line parallel with the general stratification, and are not far removed above the mass of lias limestones, which alternate with marls or shales, and constitute the lower part of the lias, taken generally, at the same place. Now these nodules contain a greater amount of carbonate of lime than the shale or marl above, beneath, or between them. We might therefore suppose that they were concentric concretions; but the fact is they are not such.

Fig. 12.

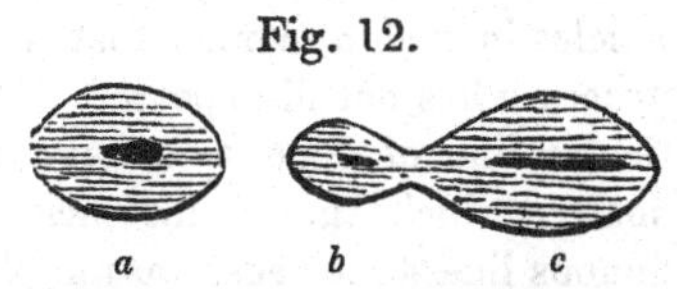

When we obtain a section of, or fracture them, we find that they are laminated, as represented above, (fig. 12.,) the laminæ having the same direction as the great mass of stratified rock of which they form a component part. The laminæ of the nodules are precisely parallel to the laminæ of the shale or marl in which they are inclosed, and little doubt can exist that they once constituted continuous portions of each other. The particles of calcareous matter have separated from the mass of marl, and have congregated together as we now see them. When we fracture these nodules through their centres, and parallel to the laminæ, we generally find some fossil, such as a fish, an ammonite, nautilus, or piece of wood, which seems to have formed the point of attraction to the various particles of calcareous matter that have assembled together. We might therefore infer that a fossil or other foreign body was always necessary to the production of the nodules; but though such bodies are very commonly discovered in the centres of the nodules, they are not always present in them. There are many without them, and some contain small scattered ammonites and other shells, not more abundant in their centres than in other portions. Hence, though fossils or other bodies may have, and most probably have, assisted in attracting these particles of matter together, they are not essential to the production of the nodules. They have resulted from the aggregation of these particles in such a manner that the original laminated structure has not disappeared. If we conceive a particular line of deposit to have contained calcareous matter which, though insufficient to constitute continuous limestone beds, was too plentiful to remain disseminated in the marl without endeavouring to arrange itself differently, and in small masses, we may probably approximate to the truth. This view is borne out by considering how some of the upper beds of the lias limestones, at the same place, are circumstanced. They are evidently only extended lines of flattened nodules, which are, however, not generally laminated, but compact, the arrangement of the particles having been more general. Fig. 13. represents such strata, or rather attempts at strata, *a*, *a*, *a* being argillaceous limestones, *b*, *b* shales.

Fig. 13.

Now the latter structure in rocks, evidently mechanically produced, and even not calcareous, must be familiar to every geologist, though it is more prevalent where calcareous matter is intermingled with other substances. It is common in various sandstones, and is evidently the result of the attraction of certain particles among each other after deposition. The lamellar structure of nodules in siliceous sandstones is sometimes also observable, an aggregation of particles having taken place on the same principle as that above noticed in the lias, the cementing matter being sometimes calcareous, at others siliceous.

Layers of nodules, not lamellar, yet composed of substances which have separated out from the constituent parts of a mechanical rock after deposition, are common in many strata. The ironstone nodules of the coal measures seem to have been thus produced. They also sometimes contain organic remains, such as portions of fossil vegetables; but as multitudes exist without organic exuviæ, they are evidently not essential to the formation of the nodules. The matter of nodules appears sometimes to have separated out from the body of the rock in which they are found in such a manner, that there has been a contraction of parts, from desiccation in the interior, producing cracks, which have subsequently been filled up, by the infiltration of carbonate of lime. The annexed sketch (fig. 14.) represents one of these concretions, com-

Fig. 14.

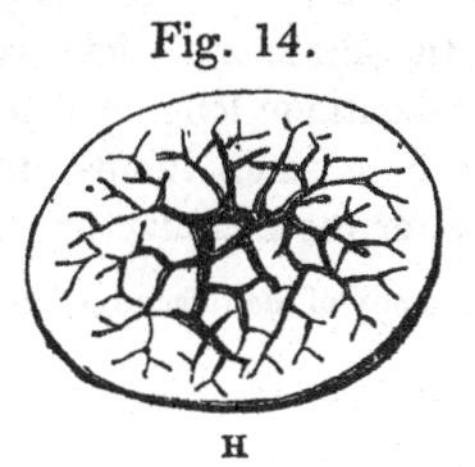

monly known as *turtle stones, ludus Helmontii*, &c. The external parts have first become consolidated, so that during the desiccation of the interior, the internal parts were compelled to separate into cracks, and shrink towards the circumference, leaving the largest fissures in the innermost parts, as must necessarily happen. The subsequent filling up of the cracks is generally illustrative of the gradual accumulation of matter from the side of such veins towards their middle portions. Coat above coat of carbonate of lime cover one another until the sides meet, highly crystalline matter filling up the irregular cavities which have often thus resulted. Nodules of this description, which generally exhibit no trace of either concentric or lamellar structure, are frequent in many clays and marls.

An aggregation of siliceous particles is pointed out by Mr. Babbage as accidentally occurring in the preparation of the clay used in the manufacture of porcelain. It may be regarded as an experiment highly illustrative of the separation, by affinity, of extremely comminuted particles of matter, deposited from water which previously contained them in mechanical suspension. In the common process of preparing the plastic substance for the potteries, flints burnt and ground are suspended with a certain proportion of clayey matter in water, and a deposit is produced of the required distribution of the particles of silica among the clay. If the compound be used in proper time, the siliceous particles remain disseminated; but if it continue long at rest, the silica becomes aggregated in small lumps, and the mass is rendered useless for the manufacture *.

* Babbage, Economy of Manufactures, 2nd Edit. p. 50.

The same author calls attention to this fact as illustrative of the formation of flints in chalk; and certainly there appears considerable analogy between them. The silica appears to have separated out of the great mass of calcareous matter, and to have arranged itself in numerous instances round various organic matters. As in the case of the calcareous aggregations above mentioned, though this frequently happens, it is not a constant fact, and we may thus consider such a nucleus not to be necessary to the production of the flint nodule. Some determining cause, though not necessarily an organic remain, must have produced the accumulation of siliceous matter in a particular point; and this cause must have extended in lines parallel to the general stratification, since the flints commonly occur as represented beneath, (fig. 15.,) *a, a, a* being lines of flints in chalk.

Fig. 15.

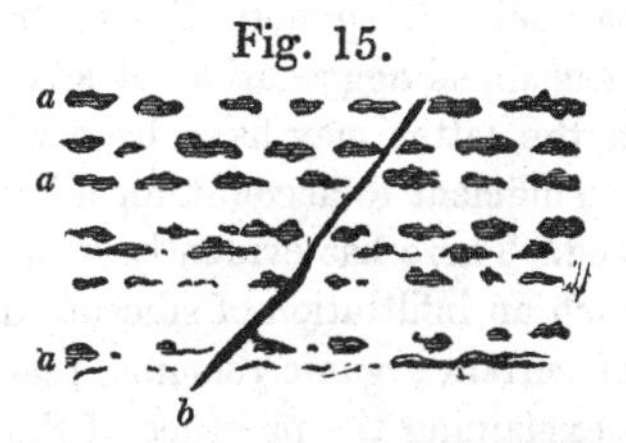

Another curious circumstance attends the separation of siliceous matter from the body of the chalk. Veins of flint traversing the lines of stratification (fig. 15. *b*), and appearing like cracks filled up by silica, are not unfrequent in many parts of the rock. Now the substance of these veins seems as good flint as any of the nodules, and we are led to conclude that they have been produced by causes similar, as far as regards the separation of the siliceous matter, to those which have formed the nodules. The regularity of the lines of flints reminds us of the regularity of lines of calcareous nodules in many clays. There was, probably, in a certain amount of deposit, formed over a certain area, a quantity of siliceous matter which had a tendency to aggregate into nodules, and the more readily as organic exuviæ constituted points of attraction. We have seen in the case of the plastic substance used in porcelain manufactures, that when we compare it with great natural operations, a short comparative time is alone necessary to cause the aggregation of the siliceous particles. Hence upon a certain amount of deposit the siliceous matter would separate out into nodules; and this operation being repeated, there would be a repetition of similar effects, and we should have lines of flints arranged in lines above each other, and parallel to the general line of stratification. The mere mechanical aggregation of siliceous particles, though the latter may have been very minute, is however insufficient to account for all the phænomena observed. There has evidently been, in numerous cases, such an infiltration of siliceous matter into the pores of certain organic remains, that there is a difficulty in explaining the presence of the silica, under such circumstances, otherwise than by considering it to have entered into these bodies in a state of solution, In certain chert nodules of the green sand, and indeed of other arenaceous rocks, we often observe a curious passage from a decidedly arenaceous structure into one apparently chemical. The interior of many of such nodules is hollow, and the mammillated chalcedonic interior has evidently been produced in the manner of stalactites. There has been a percolation of a liquid, containing silica in solution, through the arenaceous matter towards the interior of the nodule, yet the passage of one structure into the other is generally gradual. We shall have occasion to notice the flint veins under the head of veins generally.

The cleavage of rocks, which may be, and often has been, mistaken for stratification, seems due to particular arrangements of the particles of matter after the original deposition or production of the rock. It is as much observable in some of the rocks, commonly termed unstratified, as among those named stratified: in those which have resulted from fusion, as in those which have originated from solution or mechanical deposition. We should more particularly be inclined to expect cleavage in rocks whose immediate origin has been solution, and which approached the composition of a salt, such as nearly pure carbonate of lime or dolomite. Numerous limestone rocks are cleaved in directions perpendicular to the stratification, and sometimes the cleavage approaches the rhomboidal character of calcareous spar. Limestones, which even contain an abundance of organic remains, frequently present natural cleavages in such abundance that it becomes no easy matter to decide at a short distance which lines should be considered those of stratification. Attention to the general dip of the rock, and to the mode in which the organic remains occur, (beds generally containing certain marked assemblages of them,) will no doubt decide the difficulty; but there are situations where the cleavages of limestones are sufficiently embarrassing. Where the cleavage traverses organic remains themselves, dividing them in such a manner that a portion of the shell or other fossil is on each side of the fissure, the decision is necessarily easy. In some carboniferous and grauwacke limestones deceptive cleavage is very common.

Cleavage must even be more common in rocks than is made apparent in natural sections or exposures; for workmen are generally aware that the various rocks they employ for building or other purposes have what they term a *grain*, that is, the rock will split in one or two directions easier than in others. Granite-workers are, for the most part, particularly expert in finding the grain, when, to a common observer, there is no very marked difference in appearance between those parts of the granite into which they drive wedges to force open the rock into plane surfaces, and those where they say it would be useless to work. Having once ascertained the direction of the grain in a quarry, they have little trouble in raising large masses of granite, which they find readily cleavable in given directions. It is this cleavage which, by the decomposition of the granite in part, has produced the *tors*, as they are called, of Dartmoor, Devon, and which have often the appearance of art. The following sketch (fig. 16.) of part of Great Staple Tor, seen from the south-west, is more like the remains of some huge building, or

Fig. 16.

battlement, than the effect of cleavage and decomposition, which it is.

The cleavage of the Devon and Cornwall granite is frequently in given directions over considerable areas, showing that the causes producing them have acted on the large scale. The most general is from N.N.W. to S.S.E., as has been observed by Mr. Fox in the granite of Penryn*.

The cleavage of granite at its junction with slate and other rocks is sometimes highly deceptive. The granite of Devon and Cornwall has often been termed stratified from this cause. A little attention will, however, in general, show that the lines resembling those of stratification run into one another in various directions, or only continue for short distances†, in the manner represented beneath (fig. 17.).

Fig. 17.

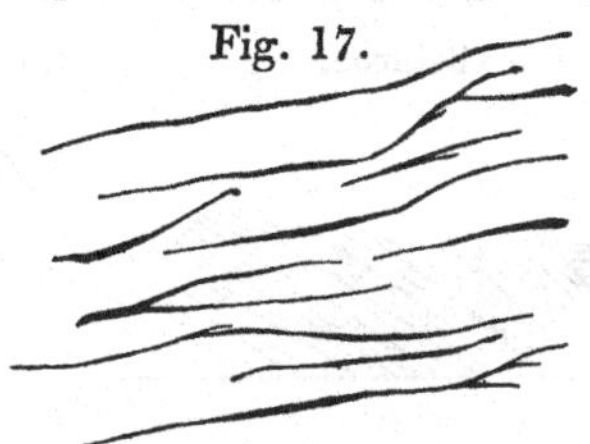

As the granite of this part of England is generally porphyritic, that is, contains large disseminated cry-

* "On the Granite District near Penryn, Cornwall," London and Edinburgh Phil. Mag., May 1833.

† The well-known Cheese Wring on the eastern edge of the mass of granite north from Liskeard, Cornwall, appears like a number of cheeses piled on one another, (whence the name,) when viewed in one direction,—therefore like the remains of beds of granite; but seen on the W. and S.W. sides is found to be one mass, with cleavage lines upon it, which do not run regularly round.

stals of felspar, part of such crystals may sometimes be observed on one, and part on the other side of the cleavage, thus clearly separating such appearances from stratification. While on the subject of these appearances in the Devon and Cornwall granites, it may be remarked, that though their posteriority to the grauwacke of the same districts may be shown from various circumstances, the cleavage, resembling stratification at a distance, runs in planes parallel to those of the stratified rock which they support; thus giving the upper surface of the granitic masses the appearance of stratified rocks upon which other stratified rocks have been deposited. The following section, on the north side of Dartmoor, will illustrate this fact.

Fig. 18.

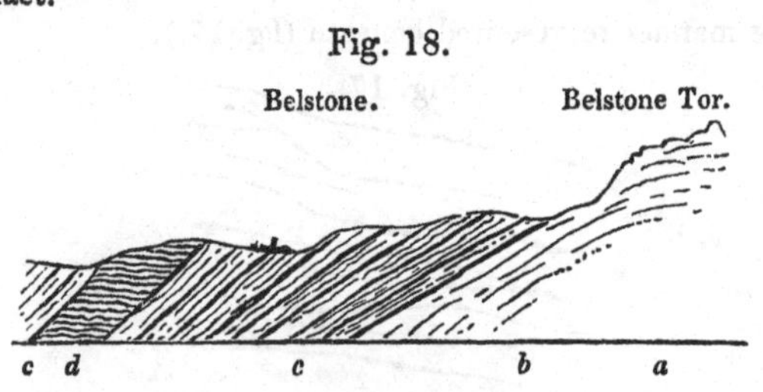

a granite, which becomes cleaved at *b* in lines parallel to the overlying altered grauwacke *c*. *d* greenstone, bounded by planes parallel to the strike or direction of the grauwacke, and thence appearing as an included bed. *c* grauwacke sandstone. *b* merely affords one view of the cleavage of the granite. It is also cleaved *perpendicularly* in lines about N.N.W. and S.S.E., and also by others which run at right angles, or nearly so, to these last, so that the intersections of the resulting planes produce large flattened prismatic blocks. The perpendicular cleavages in N.N.W. and S.S.E. lines are the most constant. The perpendicular fissures of granite on the coast of Cornwall often give a columnar character to the rock; as has been remarked by Mr. Carne.

The attraction of the particles of granite in such a way as very commonly to form either prismatic or rhomboidal masses took place apparently after the various elementary substances constituting the component minerals of granite had crystallized out, and produced the confusedly crystalline substance we now see. If this view be correct, the cleavage would be a secondary action. It might have taken place either during consolidation, or even after the mass had assumed the solid form. The cleavage of such a substance as granite would so far differ from that of regular crystals, that the latter may be considered as a definite chemical compound, while the granite is a mixture of various crystalline substances, each with its own cleavage as a separate mineral. If we consider the granite blocks to be, as has been supposed, the result of globular internal structure, we should expect that the planes, produced by the pressure of the spherical masses against each other, would not be the rhomboids or parallelopipeds we find, but other figures.

That differences of temperature may produce different arrangements in the particles of a solid body, without altering its external form, has been well established by M. Mitscherlich in his experiments on sulphate of lime and other substances. Prismatic crystals of sulphate of nickel had their internal arrangement changed in a few days by the mere action of the sun's rays upon them in a closed vessel, so that when broken they were found composed of octohedrons, with a square base, the external forms of the crystals being unchanged*. When acquainted with these and other facts of the same kind, we are led to suppose that rocks may not only become visibly altered by the long-continued action of diminished or increased heat upon them, but that they may also have their various parts differently arranged, as to mutual attraction, without the general appearance of the rock being sensibly changed. We might therefore regard the cleavage of granite as, to a certain extent, a crystalline arrangement of the mass, after general consolidation and the crystallization of the minor parts. The regularity with which granite, in various parts of the world, is split up, either naturally or artificially, into rhomboidal or prismatic masses, is quite as remarkable as the cleavage of crystals into minor portions.

Many sandstones, particularly siliceous sandstones, readily divide into flat rhombic prisms and other regular figures, which, though often observable from the decomposition of the rock, are more frequently seen when the sandstone is fractured. They are in general comparatively small and thin, such as that represented in fig. 19., which is from the grauwacke of the Hartz. The longest side of the specimen itself is five inches, and the shortest two and a half inches long, the height being about three quarters of an inch. The acute angle is = 67°, and the

* Mitscherlich, Annales de Chimie, tom. xxxvii.

Fig. 19.

obtuse = 113°; so that this portion of an arenaceous rock, bounded by cleavage planes, has all the regularity of a crystal. It is not intended to state that all the portions of detrital rocks, thus formed by cleavage into rhombic prisms, present the same angles; but the numerous cases in which there is an approximation towards them is somewhat remarkable. In many parts of the arenaceous grauwacke these appearances are common, so that beds are easily broken up into regular fragments, three, four, five and six inches in length, and less than an inch in thickness. The aggregation of particles in this case has taken place among grains of detrital matter. Nevertheless the prismatic or rhomboidal form, as the case may be, resulting from the intersection of cleavage planes, strongly reminds us, as before stated, of crystallization. Calcareous matter is sometimes found, but it is by no means a necessary condition of the rocks which thus cleave, for the same kind of cleavage is observable in sandstones in which carbonate of lime cannot be detected.

Some shales have cleavage fissures of remarkable appearances; among them we may notice perpendicular lines parallel to each other, dividing the rock in such a manner, that when the plane of stratification is viewed for some extent, the divisions have the appearance of planks ranged side by side. This kind of cleavage is well seen in part of the lias shale of Lyme Regis, beneath the Church Cliffs. The lines may be considered as imperfect developments of rhomboidal or prismatic cleavage, the cross fissures not having been formed.

Nothing is more common among argillaceous slates of various ages than the cross cleavage represented beneath (fig. 20.), *a, a, a, a, a,* being the real lines

Fig. 20.

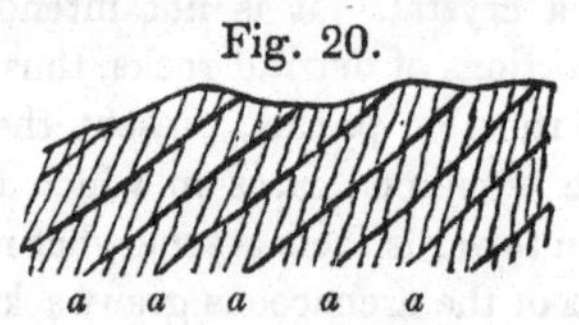

of stratification, while the other parallel lines are cleavage fissures. This structure is occasionally very embarrassing; and it is by no means easy to decide between the lines of stratification and cleavage. Such cleavage fissures must be carefully distinguished from that kind of substratification so common in many arenaceous rocks, and noticed above, which gives a diagonal arrangement to the parts of a bed. The former may be generally known by their regularity, and frequently by the considerable angle, sometimes amounting to a right angle, which the cleavage planes make with those of stratification.

This tendency in so many rocks, both chemical and mechanical, to divide into solids of more or less regularity, clearly points to some law according to which the particles of solid matter endeavour to arrange themselves, no consequence how produced, in determinate figures. Of these the prismatic and rhomboidal are most commonly observed. We have above noticed that the long continuance of some given temperature will produce a change in the arrangement of the constituent particles without altering the external form of the mass. The arenaceous hearthstone long exposed to continued heat, insufficient to fuse it, in a blast furnace at Shiffnal, and noticed by Dr. Macculloch, is a curious example of the prismatic arrangement which may, under such circumstances, take place among grains of matter accidentally thrown into juxtaposition. This fact has been adduced in illustration of the prismatic character of certain sandstones, in positions which would justify the conclusion that they also had been exposed to the long-continued action of heat, being in contact with, or near, masses of trappean rock. The disposition of sandstone to become prismatic under such conditions, which is still further illustrated by the prismatic arrangement of sandstones, baked in kilns for the purpose of hardening them, between Halifax and Huddersfield *, is highly interesting in connexion with rock cleavage. It shows that matter, in the first place mechanically aggregated, may subsequently have its particles so acted on that the mass becomes composed of numerous solids, of a certain regularity of form, while its arenaceous aspect remains.

If we admit the theory of Mr. Gregory Watt respecting the columnar structure of basalt, and rocks of that kind, we have evidence that by the pressure of spherical concretions produced during consolidation of the rock, a series of prisms are formed, hexagonal if the centres of the spheres or spheroids be equidi-

* Geological Manual, p. 472.

stant from each other; of other figures if the pressures be unequal from differences in the distances of the centres. This author supposed that, in the surface or surfaces first slowly cooled, a stratum of spheroids, with centres more or less equidistant, was produced. These spheroids, by gradual increase in their volumes, pressed against each other laterally, producing hexagonal figures when the pressure was equal. Upwards, however, supposing the bed of basalt to be nearly horizontal, they could extend indefinitely, at least to the top of the bed, if circumstances were favourable. The result would be a multitude of prisms pressed against each other, and perpendicular to the surface where the spheroids were first formed. Mr. Watt found, in his experiments on the fusion and slow cooling of basalt, that the spheroids produced were not only radiated from their centres, but were also arranged in concentric coats. He hence concluded that the joints, sometimes observable in basaltic columns, were formed by surfaces corresponding with such concentric coats. This theory is strongly supported by facts, more particularly by the horizontal arrangement of columns in perpendicular dykes, such as those at the Giant's Causeway, and other places, where the columns proceed from the sides of the dyke to the central parts, and there become confused, from the meeting of the two systems of columns, if, as Mr. Watt remarks, circumstances be so far favourable as to permit such a meeting*.

Spherical and spheroidal concentric concretions are very common in basaltic and other igneous rocks.

* Watt, "Observations on Basalt," &c., Phil. Trans., 1804; see also Geological Manual, *Art.* Unstratified Rocks, p. 469.

They are sometimes not observable until the rock has been subjected to a certain amount of decomposition, when the outer coats being decomposed, and the inner portions remaining unchanged, the rock seems composed of numerous cannon-balls and bomb-shells, separated from each other by softer substances.

Although the forms observed in these chemical rocks so far resemble those noticed above as occurring in the mechanical strata, that they are both the results of particles aggregated together after the rock has been first produced, yet they probably differ in other respects. We can scarcely conceive that the pressure of spheroids against each other would give us the flat rhombic prisms arranged as they are so commonly in the grauwacke; neither should we expect to obtain the long lines of laminæ which so frequently meet the lines of stratification at various angles in argillaceous slates. No doubt in all these cases there is an aggregation of particles giving rise to peculiar forms; but the mode in which this was effected appears to have been different.

The whole theory of altered rocks depends upon the power of the constituent particles to arrange themselves in a manner different from that in which they previously existed, and this in consequence of the continued action of a heat insufficient to produce fusion. When we see chalk converted into granular limestone on each side of a basaltic dyke, as is observable in the Isle of Raghlin on the north coast of Ireland*, we remark that there has clearly been a new arrangement of the particles of carbonate of lime.

* Buckland and Conybeare, Geol. Trans., vol. iii.; and Geological Manual, p. 279.

Instead of the loose aggregation, so well known in chalk, we find a crystalline arrangement of parts, the crystallization being more perfect in proportion to the proximity of the basaltic dyke. All analogy teaches us that in this case the alteration has been produced by heat, communicated to the chalk from the basalt. The cause and effect are apparent.

If fusion took place, there would be an end to the stratification of the rocks acted on by heat. The new arrangement of parts is effected without a liquid condition of the matter composing the rocks. That liquidity is not necessary to such changes is well proved by M. Mitscherlich's researches into those produced in the structure of some crystallized substances by different degrees of temperature. The changes caused in the sulphate of nickel have already been noticed (p. 106.). Crystals of sulphate of zinc and sulphate of magnesia, gradually heated in alcohol, lose their transparency, and are found composed of numerous small crystals, differing in form from those used in the experiment. Prismatic seleniate of zinc, exposed to the sun on paper, speedily changes into octohedral crystals with a square base. The same distinguished author observed that the optical properties of plates of sulphate of lime and other substances were altered by changes of temperature, showing an alteration in the interior structure, while no sensible exterior modification could be observed in the plates*. The various tempering of steel, and the annealing of glass, must also arise from new arrangements of the particles of steel and glass caused by heat insufficient

* Annales de Chimie, tom. xxxvii.

to produce fusion. If we take a piece of common green bottle-glass, and expose it to continued heat, insufficient to cause fusion, we obtain a crystalline substance, composed of numerous prisms arranged at right angles to the surfaces of the glass, the external form of which remains unaltered, notwithstanding the new arrangement of the internal particles.

These changes, which can readily be made apparent by experiment, prepare us for greater changes, produced by similar means, though on a far different scale, in nature. While, however, we conceive that rocks are thus frequently altered on the large scale, when in contact with masses of igneous matter which have continued to give out heat through long periods of time, we must not forget that rocks are very bad conductors of heat, and therefore we must allow a reasonable limit to an action of this kind. The well-known fact that lava currents have been traversed by men while the liquid melted rock has continued to flow in the interior, is sufficient to prove the bad conducting powers of at least such lavas. There seems, however, no exception in this respect in favour of lava, for it appears to be the case with all rocks: if it were not so, and the central heat of our globe be founded on probability, the surface of our planet would be uninhabitable by all that at present exists upon it.

The probable limits which we may assign to the alteration of rocks by long-continued heat is of necessity exceedingly difficult to determine. Experiments to illustrate this point can scarcely be satisfactorily made; for we can neither command the mass of heated matter, nor the body of rock through which the heat has to pass before all be reduced to a common temperature: neither can we command that great element, time, the necessary joint cause of so much we find on the surface of our planet.

As all rocks appear, with very slight variations, such bad conductors of heat, it follows that if a great mass of granite or trappean rock be thrown or injected among detrital rocks, such as sandstones, shales, &c., much will depend on the facility with which the heat may pass off. If a portion of the liquid heated rock be thrust through into the atmosphere, radiation will take place more rapidly than if it be entirely surrounded by rock, and therefore alteration around it would be more perfect in the latter than in the former case. Pursuing this mode of reasoning, the deeper, beneath a mass of strata, igneous rock may be injected, the greater, all other circumstances being the same, would be the alteration of the surrounding strata. The longer, also, a given heat should continue to pass through a mass of rock, the more complete should we expect to find the new crystalline arrangement of parts.

We must be careful to limit the alteration of rocks to reasonable distances, measured from the body of the heated igneous mass. The latter can only retain its heat, sufficient to produce the effect required, because the difficulty of parting with it is great. Therefore after a certain distance the quantity of heat passing in a given time becomes too small to produce an appreciable change. Hence the amount of change depends on the mass of the heated rock, the conditions, as to position, under which it may occur, and the conducting powers, as to heat, of the surrounding rocks. The first gives us the greatest amount of heat, the second the length of time it may require to pass off, and the third the distance to which it may extend with sufficient intensity to cause alteration in the particles of the constituent matter of contiguous rocks.

We shall conclude this chapter with an observation on the consolidation of rocks. It at first sight seems difficult to explain how certain sandstones become consolidated. A curious fact, noticed by M. Pouillet, seems to throw light on this subject. He states that after the plates for looking-glasses have received their last polish in the manufactories of Paris, they are placed side by side, like books slightly inclined on the shelf of a library. If they be now left undisturbed a long time, they adhere so strongly together, that not only is it difficult to separate them without breaking, but three or four pieces are sometimes also so far incorporated, that they may be worked together, and even cut with a diamond, as if the whole were originally one piece. This author also informs us that M. Clement Desormes showed him portions of plate glass, taken for the purpose of experiment from the royal manufactory of St. Gobin, which were composed of two, three, or four pieces originally distinct. "These specimens were rectangular, several inches long, and the various pieces of which they were composed, thus united by time, at common temperatures, adhered as powerfully as if formed together. Great mechanical force was requisite to make them slide on their junction surfaces; and when they were considered to have separated, it was found that, instead of sliding, they had broken, so that the junction surface of one was covered to a great extent by detached portions of another*."

We might, perhaps, infer from the facts above noticed, that grains of quartz, so commonly constituting sandstones, may be permanently made to cohere by long-continued pressure, under favourable circumstances, beneath great masses of rock. The pressure to which many arenaceous rocks have been subjected must have been enormous; and it is remarkable that slightly aggregated sandstones are more common among the more modern than among the more ancient strata. There can be no doubt that cementing matter, chemically produced, has greatly promoted consolidation; yet, probably, long-continued juxtaposition, under pressure, has not a little aided the process. We certainly discover sands without cohesion among strata, but such are very rare among the older rocks.

It may be here observed, that arenaceous rocks long exposed to the percolation of water, charged with foreign matter, would have a tendency to be consolidated by the deposition of such matter among the constituent grains. That silica in solution percolates through such rocks, we have proof in the fossil shells often converted into chalcedony in sandstones, the calcareous matter of the shell having been removed, and silica infiltrated into the mould thus prepared to receive it. That silica, carbonate of lime and other substances percolate through rocks of much finer tex-

* Pouillet, Elémens de Physique Expérimentale, tom. iii. p. 41. Seconde Edition. Paris 1832.

ture than sandstones, is proved by the formation of agates and other minerals in the cavities of basaltic and trappean rocks, such cavities having originally been produced by bubbles of gas when the rock was in a state of fusion.

CHAPTER VI.

We have previously remarked that the surface of dry land has been shattered and broken into fragments, and that there is rarely an area of a few square miles which does not bear marks of having been acted on by disrupting forces. Seeing that this is the case with such portions of the earth's mineral crust which rise through the ocean into the atmosphere, we may infer that the solid matter beneath the ocean is not exempt from similar fractures; for we cannot suppose that the dislocations of the earth's surface have been confined to those portions only which are accessible to our observation. We may, indeed, consider that the whole mineral crust of the globe is broken into fragments of various sizes, forced into contact by gravitation towards the centre of the earth. The more recent rocks must necessarily be less fractured as a whole, than those which are more ancient, also taken as a whole; for as the former rest on the latter, dislocations in them caused by forces acting from beneath would traverse both equally, while the more ancient rocks may have been, and no doubt have been, broken before the comparatively recent rocks were formed. It would therefore follow, that in a district composed of more recent rocks, the fractures visible in it would not afford a just estimate of the amount of dislocation to which that particular part of the earth's mineral crust has been subjected; for the strata beneath may have been greatly broken before the rocks visible in the district were in existence. Consequently, in tracing lines of dislocation, we should be careful to observe, whether they terminate at rocks newer than those in which we had traced them, whether they are continued in any older rock beyond the area of the newer rock, or whether they have in all probability never extended beyond the point noticed, the natural fracture having there terminated. In the first case we should have evidence that the particular line of dislocation was produced before the formation of the newer rock; in the second, the relative date would be uncertain.

That disturbing forces have acted on rocks after their formation, is as well shown by tilted and contorted strata as by fractures. Extensive ranges of country often have the beds of rock of which they are composed thrown into particular lines of direction. This fact is frequently observable in mountain-chains, though it is by no means confined to them. Such lines, when considered in the usual manner with reference to our general ideas of distance, appear of considerable length; but when viewed, as they should be, in connexion with the whole superficies of our spheroid, a large proportion of them lose their apparent importance. Many of them are then readily seen to be so short, that the cracks or elevations of strata by which they are marked may readily be conceived to have been effected by comparatively small intensities of force. It is perhaps to a want of due attention to the relative proportions of the radius of the earth to the height of mountains, and of the length of mountain-chains to the superficies of the globe, that those who have considered such lines of tilted strata or dislocations as the result of a few, instead of very numerous elevations, have been supposed to call in the agency of forces so tremendous as almost to alarm the understanding; while they really have recourse to forces comparatively insignificant.

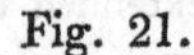

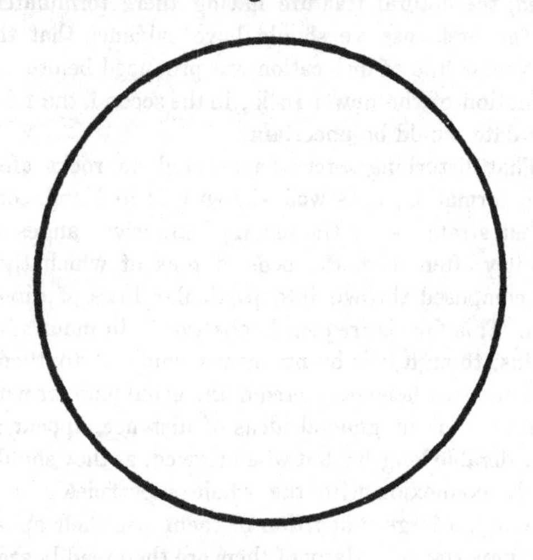

The annexed diagram (fig. 21.) may afford an idea of the proportion of the crust to the diameter of the earth. Considering the whole circle as a section of our planet, the black outer line would be one hundred miles thick, that is, the thickness of that line bears the same proportion to the area and diameter of the circle in the annexed figure, as a crust, one hundred miles deep, would bear to a real section of the earth made through its centre*. It is only necessary to state that the highest mountains on the face of the globe do not attain an elevation of six miles, to perceive that inequalities of such an amount, even if they were common, instead of being, as they are, rare, might be produced either by contractions or expansions in the mass of the globe itself, and the necessary fractures and dislocations be comparatively insignificant.

If the theory of central heat be founded on probability, as we conceive it to be, the very general occurrence of tilted and fractured rocks is of easy explanation. We require a general cause for the production of so general an effect. If we suppose with M. Élie de Beaumont that the state of our globe is such that, in a given time, the temperature of the interior is lowered by a much greater quantity than that on its surface, the solid crust would break up to accommodate itself to the internal mass; almost imperceptibly when time and the mass of the earth are taken into account, but by considerable dislocations according to our general ideas on such subjects. We should expect to find the lines of fracture innumerable, precisely as we do find them. We should also be led to anticipate that new dislocations would be more readily effected through old lines of fracture, and that under favourable circumstances, broken and tilted masses would be thrust up into ridges or mountain-chains. If, in the annexed exaggerated diagram,

* For a figure exhibiting on a larger scale the relative proportions of the Alps, Andes and Himalayan mountains to the radius of the earth, see my Sections and Views illustrative of Geological Phænomena, pl. 40.

(fig. 22.), the outer circle represent the crust of the earth at a given time, and that a subsequent contrac-

Fig. 22.

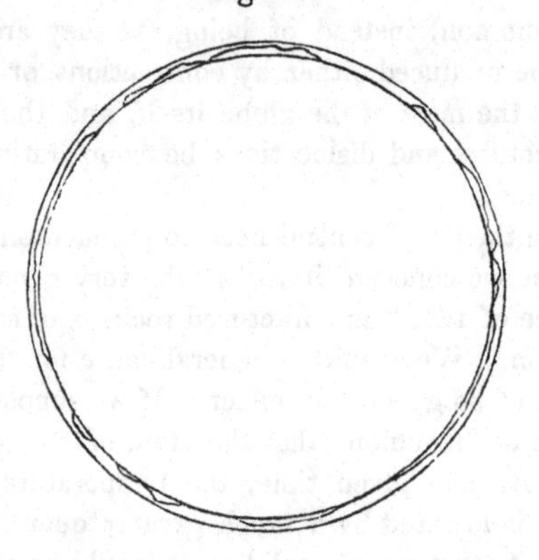

tion of the internal mass, which does not produce a proportional effect on the previously existing crust, requires the latter to descend to the inner circle, dislocations and contortions of the strata would take place, and masses would be squeezed up against each other in order to accommodate the old surface to its new condition. A sectional line would be produced, something similar to that represented (fig. 22.) between the two circles, though by no means bearing a like proportion to the diameter of the inner circle; for it would require a circle many inches in diameter to render such inequalities visible and proportional at the same time.

If the edges of a considerable crack or fissure be acted on in the manner here supposed, they would apparently be forced upwards, by lateral pressure, with respect to the sea-level, which might be depressed, filling the new hollows. When we estimate the solid matter required to produce a very considerable addition to a mountain-chain, we find that an equal volume of water received into a corresponding depression of surface, would, if distributed over the mass of the ocean, cause far less general depression of level than we might anticipate.

Some geologists, misled by the idea that an accumulation of small forces will always produce the same effect as the sudden exertion of a great power, equal in intensity to the sum of the smaller forces, deride those who conceive mountains to have been raised by greater intensities of force than those observable in a modern earthquake. This is as if, having piles to drive into the ground, which it required the exertion of considerable force to pierce, we employed a pile-driver with the necessary power, and should for so doing be laughed at by our friends, and advised not to be thus extravagant in machinery, but to perch a man with a hammer on each pile, who, by repeatedly striking small blows on the head of it, would, it was considered, accomplish the end equally well, and at much less cost.

It is anything but desirable to have constant recourse to comparatively great power in explaining geological phænomena, when the exertion of a small force, or of an accumulation of small forces, will afford sufficient explanation of the facts observed. This, to follow up the above illustration, is as if we were determined never to suffer a stake to be driven into a hedge without a pile-driver. We should be careful not to suffer any preconceived opinion to bias our views either way. We may be wrong in our endeavours to attribute the exertion of forces necessary to produce mountains, fractures and great disturbances of rocks to any particular causes, but the probable intensities of the forces themselves should be most carefully estimated, so that should we fail in the first, we may approximate towards truth in the second case, and thus render our investigations of some use to the progress of science.

In our endeavours to account for mountain ranges, with reference to the dislocations and disturbances of rocks so commonly observed in them, circumstances which have occurred beneath have always been considered as the cause of their existence. Even in Mr. Lyell's very ingenious theory respecting variations in climate produced by differences in the elevation and position of land, the supposed changes in the relative position of solid surface-matter are necessarily caused by something acting beneath it. Consequently the climates so produced must be secondary, and depend for their existence on a previous movement of the earth's solid surface. Now as like effects must be produced by like causes, and mountains are not local, but distributed over the surface of the world, (for we may fairly consider that they exist as well beneath the ocean as on the dry land, indeed islands are often little else than the exposed tops of subaqueous mountains,) it follows that the cause producing them, together with the great dislocations and disturbances of rocks observable on dry land, is common to at least the surface of a spheroid placed immediately beneath the crust of the globe.

What that cause may be is a separate question, and is one to which we now propose to call the attention of the reader.

That the relative level of sea and land can be changed by earthquakes seems a well-established fact; and it is more particularly to the labours of Mr. Lyell that we are indebted for an accumulation of a mass of evidence on this subject. It is only when Mr. Lyell limits the intensity of this action to the ordinary force exhibited in a modern earthquake that we are inclined to differ from him on this head. Earthquakes are occasionally felt, though with variable degrees of intensity, in all the known parts of the world. Like volcanos, they may be considered the result of general causes existing beneath the surface of the earth's mineral crust. Now earthquakes produce long cracks and dislocations of strata perfectly analogous to faults. The most remarkable ridge produced by an earthquake, and hitherto brought to light, is that noticed by Captain Burnes, as extending a distance of fifty miles in an east and west direction across the delta of the Indus, with a breadth, in some places, of sixteen miles, and an elevation of about ten feet above the level of the delta. This effect was produced by the Cutch earthquake, as it is termed, of 1819. Here we have a miniature representation of a mountain-chain, so far as the elevation of land along a given line is concerned. Fractures are not noticed, but we can readily conceive that they would be produced by the further exertion of an analogous force. A repetition of the same intensity of force, acting along the same line, would, if the beds beneath were sufficiently consolidated and brittle, cause dislocations; if they were soft or yielding, they would be contorted, or drawn out and merely bent, as the case might be.

If fractures be once formed in the crust of the earth, whether elevation be caused by squeezing the edges of the fracture against each other, or by matter struggling to free itself, the crust would be more readily acted on in the lines of previous fracture than in any others, for they would also be the lines of least resistance. It would, therefore, be no matter of surprise that many successive earthquakes, as we might term them, should act in the same line, or rather that their greatest effects should be produced in such lines. Now if mountain ranges bore no marks of the exertion of greater intensities of force than those exhibited in modern earthquakes, the mere elevation of their masses to comparatively considerable heights above the level of the sea could apparently be produced as well by an accumulation of the effects of such forces as by any other means. The examination of mountains shows us, however, that we must have recourse to greater intensities of force to account for the effects we there see produced.

It will be readily granted that in many faults or dislocations of strata it would be exceedingly difficult to say whether the amount of dislocation (and it may be considerable,) has been produced by repeated shocks, or by one of a more powerful kind.

Fig. 23.

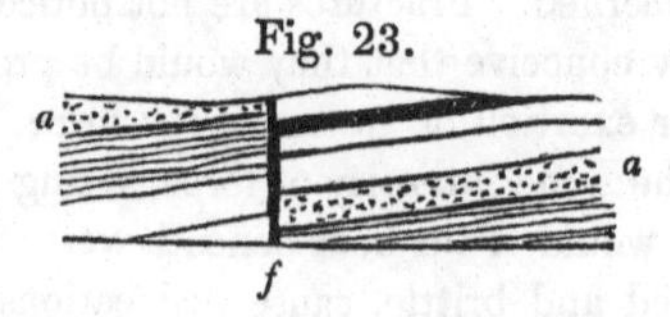

If *f* (fig. 23.) be a fault traversing strata, and the bed *a* on the one side be removed, either by elevation or depression, to *a* on the other, it would be difficult to say, from this section alone, whether the difference in the levels of the once continuous bed *a* be caused by several shocks, or by one. The polished surfaces, so common in faults, will not assist us much in the determination, as one powerful squeeze would produce the same kind of polish as many minor movements.

Neglecting for the present many dislocations which there is great difficulty in considering produced otherwise than at one snap or break, let us consider those contortions of strata so common in mountain ranges, and not very unfrequent elsewhere. These exhibit not only the effects of force, but also some information as to its intensity. Contortion requires that the rocks in which it is observable should have been in a yielding state, and that the particles were capable of a certain movement among each other, so that when force was applied no absolute fracture was occasioned. Sir James Hall has long since shown that to produce contortion by lateral pressure, there must be resistance both above and beneath, the former at least being capable of yielding in a minor degree. He illustrated this fact by experiment, and showed by means of a diagram, of which the annexed figure (fig. 24.) is a sketch, that these conditions are necessary to the production of contortions by lateral forces. Sir James Hall took various pieces of cloth, some linen, some woollen, and placing them horizontally on a table (*c*), covered them by a weight (*a*) acting horizontally on the pieces of cloth. He then ap-

Fig. 24.

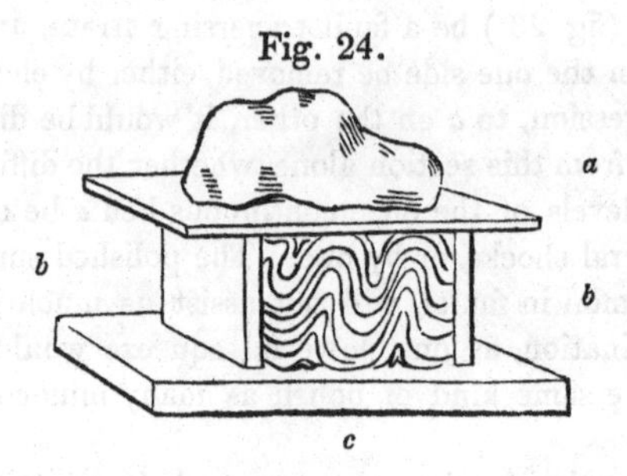

plied forces laterally (*b b*), and found that while the superincumbent weight (*a*) was raised to a certain extent, the cloth was folded and contorted in a manner perfectly analogous to the contortions of rock observable in nature*.

It will be obvious that if there be resistance only on one side, and sufficient pressure above, contortions will be produced. It also follows, that if a mass of solid matter be thrust between yielding beds, and that they cannot move sufficiently upwards, they will be squeezed and contorted so as to permit the presence of the intruded mass. Now it is worthy of attention that contorted strata are common on the skirts or flanks of many mountain-chains, appearing to show that, before the latter attained their existing forms, there was a pressure from the central parts outwards, causing the lateral contortions we observe.

* Transactions of the Royal Society of Edinburgh, vol. vii. p. 85. Sir James Hall subsequently invented an instrument to contort clay, laid in miniature beds; the principle, however, is precisely the same, and the original experiment not only satisfactory, but so simple that any one may make it on a table with a few books and pieces of differently coloured cloth.

Fig. 25.

The above (fig. 25.) is a section of the Alps, by Dr. Lusser, from the well-known Rigi to the Hospice of St. Gothard; *g*, being gneiss; *l l*, limestone; *n*, nagelfluh or conglomerate. It is not exaggerated, but affords a fair view of the manner in which the calcareous Alps are contorted on the flanks of the central range. To produce this effect, we seem compelled to suppose the whole mass of the calcareous Alps (a series of mixed strata of limestone, argillaceous slates, shales and sandstones, the former predominating,) to have been in a yielding or comparatively soft state. We can scarcely suppose, with any approach to probabilities, that the soft yielding condition of this mass of matter should have continued sufficiently long to enable a succession of small shocks, of no greater intensity than those of a modern earthquake, to have acted upon it. The whole strongly impresses us with the idea of a powerful exertion of force, squeezing the limestone and associated beds outwards. And, as if to support this view, the nagelfluh or conglomerate of the Rigi, a rock really posterior to the limestone series, is tossed over; so that it plunges beneath the latter, and appears to support that which in reality supported it before the exertion of the force that upset the whole. This section is so striking that it may be supposed to be selected from a multitude of others on which a contrary opinion might be founded. It certainly is a good example of contorted strata on the large scale; but numerous others, equally good, might readily be adduced. In the Alps themselves better sections might be produced of huge masses of matter, constituting whole mountains, fractured and tossed over in such a manner that it seems impossible to consider them otherwise than broken and heaved out of their original places by the exertion of powerful forces; forces of an intensity in comparison with which that of a modern earthquake would be trifling. It must not hence be concluded that we consider the powerful exertion of similar forces to be now impossible, but simply that the intensity of force exhibited in a modern earthquake is insignificant when compared with those which have, and may hereafter produce, high ranges of mountains.

The amount of lateral pressure and resistance to it are not the only circumstances to which we must direct our attention; we have also to account for the necessary soft condition of the rocks acted on, and the superincumbent weight by which they are prevented from being merely thrust upwards. If, in the annexed section (fig. 26.), *a a* represent two rocks,

Fig. 26.

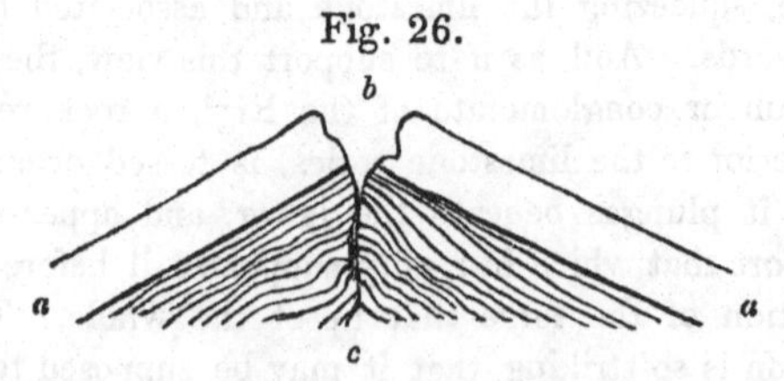

or fractured portions of one rock, squeezed together,

in consequence of depressions at *a* and *a*, they will be merely thrust upwards at *b*, with perhaps a small amount of contortion of the softer beds at *c*, if there be such in the lower parts, and therefore not only exposed to lateral pressure, but to that arising from the weight of the superincumbent mass of rock. If, instead of being squeezed together by depression on either side, the rocks *a a* be thrust upwards by the intrusion of a mass of rocky matter, the section would be as annexed (fig. 27.), where *b* represents the in-

Fig. 27.

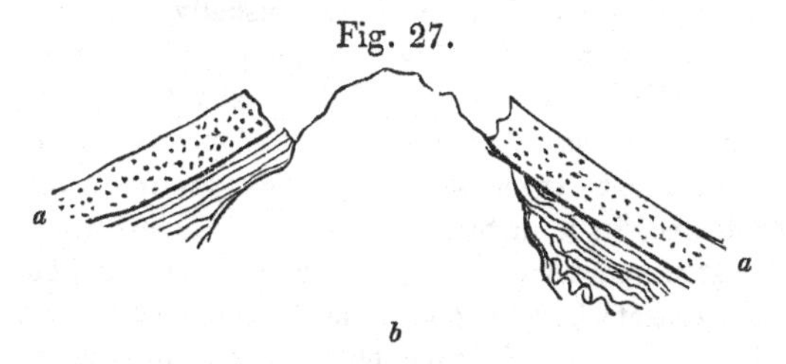

truded matter. *a a* would be merely tilted, with perhaps some contortion of the softer inferior beds, if the latter be more yielding than the mass intruded.

Neither of these hypothetical sections agrees with that of the northern side of the Alps previously given (fig. 25.). We can only obtain the observed contortions beneath pressure, and that of no ordinary kind. We might, indeed, get contortions towards the central parts in these sections, when the weight of the superincumbent matter was sufficiently great: but to have produced such an effect in the Alps, we must suppose the subsequent removal of such an enormous mass of superincumbent matter, necessary to cause the effects observed, that we are quite at a loss to conceive how this has been accomplished. An immense volume of matter has indeed been carried away, as the great masses of conglomerate and sandstone on the flanks of the chain, and evidently derived from it, abundantly testify; but how far the amount of this mass would be sufficient for the effects required is not apparent.

The question of how such a volume of matter as that composing the contorted calcareous Alps was sufficiently softened, if previously indurated, or kept soft, if never consolidated as rocks usually are, before acted on by the contorting cause, appears to be one of no ordinary difficulty; and this difficulty not only attends this case, but also an abundance of others. As we might expect, shales and clays are frequently contorted, in a disturbed district, while associated limestones and sandstones are merely fractured. The harder strata have snapped, and the softer have bent. There is nothing in such phænomena which is not readily understood. When, however, we find hard siliceous and other sandstones, compact limestones, and even brittle substances, twisted and bent into every imaginable shape, and requiring a very soft yet tenacious condition of matter, the subject is anything but easy. If we have recourse to heat, we must recollect that rocks are very bad conductors of it, and, consequently, that no great thickness of them could be softened, so as to submit to flexure from this cause without fusion in the lower parts; at least such would be the effect we should anticipate from our present knowledge on such subjects.

When we suppose that contorted rocks were not consolidated before they took the forms we now see, there is again difficulty, more particularly when we consider that rocks of different ages are sometimes twisted together. Compact limestones, so frequently contorted, would scarcely remain long unconsolidated if they have been produced by a chemical deposit of carbonate of lime, either alone or among organic remains. Modern deposits of carbonate of lime, whether formed in the atmosphere or beneath water, either fresh or of the sea, are indurated almost from the commencement of the deposition. So that if the limestones of former periods have been produced in a manner in any respect analogous, many must have been indurated before they were contorted. Carboniferous limestone and coal-measures are twisted together, evidently by the same disturbing cause, in Pembrokeshire and other places. Now, in this case, we can scarcely conceive the carboniferous limestone otherwise than consolidated prior to its present condition as regards disturbance; for the coal-measures, a deposit clearly requiring a great lapse of time for its production, were formed before both rocks suffered equally from the same contorting force.

One thing we certainly are unacquainted with, and which may assist us in the explanation of these and similar facts, and that is, the extent to which the particles of various rocks, particularly when water is disseminated among them, may be compelled to move among each other by great lateral pressure, and beneath a superincumbent weight, so great that they cannot absolutely separate from each other. The resistance would no doubt be enormous, so great indeed that we should expect the pressure would tend rather to pulverise than contort the mass, if solids be forced against solids. It would not, however, be dry; it would be variously mixed, according to the constituent rocks, with water; for the latter is disseminated among strata to a greater extent than is generally imagined, as may be readily shown by subjecting portions taken from considerable depths to immediate experiment. Water is even disseminated among limestones; among the less compact it must exist in considerable quantity taken as mass; for when dry specimens are placed in contact with water, capillary attraction is as apparent as if a lump of sugar were used instead of the limestone. Dry oolite, such as that of Bath or Portland, absorbs water with considerable rapidity when placed in contact with it. It is exceedingly difficult to expel moisture from rocks, particularly if the fragments employed be large; but if moderately sized specimens of mixed rocks, more especially of the superior stratified or fossiliferous strata, be taken and exposed to considerable heat, sufficient to drive off at least a large part of the contained water, then weighed, and afterwards allowed to absorb as much water as they can take up, and again weighed, the difference of weight will be found greater than at first sight might be anticipated.

Although water is thus widely disseminated among rocks, it does not follow that its presence there is alone sufficient to give such softness to rocks that they may be twisted and bent in any direction. If it were so, rocks, containing a mean amount of disseminated moisture, would generally be bent, not fractured, when acted on by powerful forces. It may, however, assist contortion by permitting a certain amount of passage among the particles of the bodies acted on, and also by affording a certain kind of ductility to their mass. That the particles of solid matter may be made to pass more freely among each other than was once considered possible, and without such matter becoming liquid, is proved by their change of position in cases previously noticed (p. 112.), where their relative positions must have been altered in a most curious manner, even in one case, as it would appear, by mere exposure to the solar rays. The nearly constant presence of disseminated water among rocks, under ordinary circumstances, is here brought forward for the purpose of exciting further inquiry, and as one of the elements that should be taken into account when we study the causes of contorted strata. We may here make another remark. Numerous contortions are often observable among the stratified rocks where igneous rocks are also common, and where the latter seem evidently to have been protruded at the same time that the contortions were effected. The mere lateral pressure of liquid melted rock against a solid stratified mass would not contort the latter to any considerable extent, if the fluid rose freely upwards into air or water, encountering little resistance in that direction. It would no doubt strive to introduce itself among the strata, and would effect this according to the height and consequent pressure of a superincumbent column of matter, as has been noticed elsewhere*; but this would be a very different operation from squeezing up the whole mass on either side into great bends and flexures of every description.

Now the association of igneous and contorted rocks

* Geological Manual, p. 140.

is so common, that there must be something connected with the former, which has had an influence on the latter. A liquid mass of fused rock cannot be forced from beneath through solid stratified masses without the prior fracture of such masses, caused either by the force with which the fluid matter is propelled upwards, the escape of vapour or gas from a highly compressed state, the subsidence of two ends of a large tabular mass of rock, by which a rent or dislocation is produced in the central parts, or by various combinations of these disrupting forces. If, however, a mass of liquid rock would not by mere lateral pressure cause the great contortions we observe, its heat would tend to drive the moisture in the stratified rocks to a greater distance from the fused mass. There would, in fact, be a tendency to steam the stratified rock, and the heated aqueous vapour would pass more readily between the laminæ, or among the fissures of stratification and cleavage, than in any other direction, for they would be the lines of least resistance. Hence a rock may, under such conditions, be heated beyond the part that would be acted on if it were dry, and a superabundant amount of moisture be forced into those portions which retained a comparatively low temperature. A state of things would be produced by no means unfavourable to contortion, and not without its influence in the chemical alteration of rocks, in cases where igneous matter is protruded through, or injected among them *.

* It will probably assist us in considering the quantity of water which *can* percolate through rocks in a short time, if it be recollected that, from the returns of duty done by the steam engines in the mining district of Cornwall, the mean amount of water

The most favourable condition for contortion appears to be pressure of solid matter on yielding stratified substances, which while they bend, also slide to a certain extent on the planes of stratification. Stratified rocks do not constitute one homogeneous solid

raised by the Cornish steam engines in 1833 was equal to 16,550 imperial gallons per minute, and that of this amount the Consolidated mines alone afforded a mean of 2161 gallons per minute.

My friend Mr. Richard Taylor informs me that in all the mining districts where he has made observations, (and his opportunities for so doing have been abundant,) he found the quantity of water in a mine to vary with, and depend upon, the rain falling in the districts where it was situated. Seasons of heavy rain are followed by a great increase of water in a mine, and the reverse happens from drought. The time which elapses before the effect of these causes manifests itself in the mines varies considerably, and depends on the geological character and physical features of the country where the mines occur, and even on the peculiar geological circumstances in each mine in the same district. He has observed that in carboniferous limestone districts the percolation of rain-water is particularly rapid, and such that a heavy fall of rain will overpower the machinery sufficient for ordinary drainage. He has known instances where a heavy fall of rain has produced within three or four hours so great an increase in the water of two mines, Pent y Buarth and Cathole, on the Mold mountain, Flintshire, that the steam engines, after having been increased in speed from 3 or 4 strokes per minute, to the extent of their power, about 5 times as many, were yet incapable of preventing the water from rapidly increasing and filling the workings. It appears that the physical features of the country favour the introduction of rain-water into these veins, and that there are great cross courses or faults which can be traced several miles, and which pour torrents of water into the veins. They are, in fact, channels for subterranean rivers, and Mr. Taylor remarks that the latter carry with them a quantity of sand and gravel, the detritus of the rocks through which they pass.

The collective amount of water which passes through faults and fissures must be enormous, and when added to that contained in rocks themselves, shows us the volume of it which may percolate downwards. Probably there are compensations, to a certain amount, in the proportions of water falling on the surface and percolating at moderate depths; for my friend Mr. Davies Gilbert calculates that a fall of rain, one inch deep, upon a statute acre is equal to about 100 tons.

mass even in cases, which are necessarily rare, where the strata are all similar in every respect: they are masses composed of numerous solids piled on each other, and readily separating in the lines of the beds. If the particles composing the strata can move among each other to a moderate extent, that is, the rock be sufficiently softened, and lateral pressure applied, they must become twisted and bent under a superincumbent weight of sufficient power. We can readily conceive, in the section previously given (p. 129.), that the matter of the calcareous Alps has been squeezed laterally by the pressure of the gneiss and other rocks of the central chain upon it. This would equally happen, whether the central Alps were thrown up by pressure from subsidence on either side, or by powerful expansive forces acting from beneath, and overcoming the resistance above. In many cases we may suppose the strata unconsolidated prior to contortion; in others, they appear to have been softened after consolidation. Heat and moisture would have great influence on the condition of the stratified matter so softened. Contortion from lateral pressure cannot be produced without a superincumbent weight, which, though yielding, is sufficient to prevent the mass from being merely turned up. When, therefore, we find contorted strata on the surface of dry land, we know that they are not now in the same relative position in which they were so contorted. The pressure of the atmosphere being insufficient to produce the effect required, the essential superincumbent weight, no consequence what that weight may have been, is removed.

CHAPTER VII.

We have seen that tilted strata, forming mountain ranges and many other inequalities on the earth's surface, may be due to two causes. They may be caused by the upward pressure of the sides of a fissure against each other, arising from the unequal subsidence of large masses, or by the intrusion of matter thrown up by elastic forces acting from beneath. Radiation of heat from the mass of the earth would necessarily decrease its volume, from the consequent approximation of particles; whereas, if we suppose the oxidation of a metallic nucleus, having the same temperature as the surrounding planetary spaces, the volume of such nucleus would be increased. It would, in the latter case, take oxygen from its liquid or gaseous envelope, and consequently abstract matter from it; the surface of the metallic nucleus would also become heated from the union of oxygen with certain metallic bases, but the expansion thus caused would gradually subside, and leave only the increased volume of solid matter arising from the new combination.

A cold metallic nucleus, particularly if sodium, potassium, and bodies of the like kind, were abundant on its surface, would speedily become oxidized, whether covered by water or by an atmosphere containing oxygen. But the oxidation of the surface being accomplished, and admitting that numerous cracks and fissures might be produced in the first crust, from the percolation of water through it, the oxygen combining with metallic bases and the elastic force of the disengaged hydrogen forcing it up; one does not exactly see how those long lines of elevation were produced, that are so common on the earth's surface, and of which the most remarkable appear to be the mountain ranges of North and South America. The heat caused by the combination of oxygen with such substances as potassium and sodium would not only assist in producing fractures by giving greater expansive force to the hydrogen, upon the supposition of water percolating to such bodies, but it would often cause the fusion of the inferior portion of the previously oxidized crust, which might thus be forced upwards through the fractures.

Now gaseous matter is evolved, and liquid melted rock forced up in so many points on the earth's surface, under circumstances which render the theory of the percolation of water to the metallic bases of certain substances so plausible, that it is one we should not neglect; neither should we reject it because it does not afford a good explanation of many phænomena which have been attributed to combinations of this kind. We should rather seek to discover the extent to which it may account for observed phænomena when combined with the theory of central heat.

It has previously been observed that the gaseous condition of our planet, caused by sufficent heat, might produce an interior metallic crust above which the greater proportion of oxygen and some other simple nonmetallic substances might be driven, thus leaving the metals, in the lower part of such crust, in a great degree uncombined with other substances. We will not again repeat what may be considered the probable effects of radiated heat, both on the mass and surface of the globe. It is only necessary to remark, that the fissures and dislocations so produced would be highly favourable to the introduction of water to the metallic bases of certain of the earths and alkalies. By combining the effects that might be caused by the radiation of central heat, and by the percolation of water to the metallic bases of certain earths and alkalies beneath a given depth of the earth's oxidized crust, we seem to arrive at explanations more or less plausible, which are not so readily attained by means of either theory taken by itself. The great leading elevations of land, observable in our continents and islands, as also the principal lines of disturbed strata, seem best explained by the gradual refrigeration of the mass of matter composing our globe, while a large proportion of phænomena observable in volcanic eruptions appear best to accord with the theory of the percolation of water, containing certain substances in solution, to the metallic bases of the earths and alkalies. We should in this, as in many other cases, strive rather to combine fair explanations of phænomena, though they may at first sight appear somewhat discordant, than to adopt one leading theory with the determination to see nothing except through its medium, and hence distort, and even omit, facts, often unintentionally, and in the heat of contest, for the purpose of supporting it.

It has often been observed that thermal springs are thrown out from fractured or disturbed strata, either in mountain-chains or in more level lands, and that the constancy of their temperatures is very remarkable, the cases being exceedingly rare in which any change has been suspected. Such effects can, we should conceive, be only produced by some general like causes. General heat beneath the surface, arising either from central heat, or from a decomposition of part of the water percolating to certain metallic bases, supposing these to be commonly distributed beneath the earth's oxidized crust, would be readily obtained. There is, however, this difference between the two causes; in the former the water would be little more than heated, while in the latter a part of it at least would be decomposed, liberating a body of hydrogen, which must either combine with something beneath, or eventually pass upwards into the atmosphere. Sulphuretted hydrogen, a compound very readily absorbed by water, is indeed contained in some thermal springs, but it does not occur in all. Many thermal springs neither give us hydrogen combined or uncombined, except as a component part of the water itself. If, therefore, such thermal springs have been heated by the decomposition of part of the water percolating to certain metallic bases, the hydrogen of the water decomposed must either remain behind, combined with some substance, or be set free. We can scarcely consider the latter probable, for the vent which would permit the undecomposed water to rise to the surface of land, would permit the escape of the gas; and as to the former, there are many difficulties.

Should water percolate to the metallic bases so often mentioned, we should expect that the resulting action was more violent in its effects than the general phænomena of thermal springs seem to attest; something more resembling volcanic action. The effects productive of thermal springs may indeed be secondary, that is, the heated water and the contained substances of many springs may be driven off in the shape of vapour and gas, the aqueous vapour becoming liquefied, and absorbing certain gaseous substances, such as sulphuretted hydrogen and carbonic acid, when the fissures of the rock through which it passed were cool enough to permit its liquid state, and the temperature sufficiently low. This seems to be actually the case where condensed aqueous vapour, containing gases which it has absorbed, flows out of clefts of volcanos in the shape of hot springs. The steam rushing out of the Solfatara, near Naples, was found by Dr. Daubeny to contain sulphuretted hydrogen and muriatic acid. Now it is obvious that if this steam passed upwards through a fissure of rock, sufficiently cool to condense the aqueous vapour, the result would be a thermal spring; the elasticity of the steam beneath being sufficient to propel the water upwards and produce a flow of it, the requisite additional quantity being always furnished by the condensation of the steam continually discharged. It would moreover be mineral, as it is termed, more particularly if the muriatic acid united with some substance, such as lime or magnesia, and thereby produced a compound soluble in water, as might readily happen in the passage of the water through limestones, either common or magnesian.

It will be obvious that many other substances, besides those noticed above, might be mixed with the steam rushing out of the Neapolitan Solfatara, and very probably do accompany the aqueous vapours discharged from many other solfataras. The results of such causes would be the production of mineral and thermal springs in abundance, when the steam was condensed in clefts or fissures of rock, sufficiently cool. So long as there was a supply of steam beneath, its elasticity would probably cause the propulsion of the condensed aqueous vapour above it: this cause would also explain the constant temperature, as regards each, of thermal springs. While the steam in a cleft rose to a given height, so long would the temperature of the water discharged on the surface remain the same, being exposed to equal circumstances. From the bad conducting power of rocks, the sides of the cleft or fissure would soon take their proper temperature, which they might keep during centuries.

Dr. Daubeny, in his remarks on thermal waters, has insisted on the very common presence of nitrogen in them as a proof that the water has originally been derived from the surface of the globe, that it there contained atmospheric air, and that, descending into the earth, the atmospheric air was deprived of its oxygen by some process of combustion. We have seen (p. 11.), that nitrogen is not altogether absent from rocks, being contained somewhat abundantly in coal. In this case, however, it may be regarded as existing in a secondary condition, having been derived from the atmosphere in the first instance. Nitrogen has not been detected in rocks which are not fossiliferous: it does not indeed follow that this substance may not exist as well beneath the crust of the earth as in the atmosphere; but taking our present knowledge as a guide, the probabilities are in favour of its being confined to the atmosphere, or to bodies which have acquired it from the atmosphere in the first instance. If thermal springs were confined to districts of coal-measures, or other rocks in which nitrogen was entombed, or to situations where such rocks may occur beneath those strata whence the springs actually rise to the surface of land; the presence of this substance might be due to its escape from the rocks traversed by the waters in their passage upwards. The occurrence of nitrogen in the thermal waters of numerous localities, where we have no reason to suspect the existence of similar rocks, or those which might afford this substance, incline us to believe that, contained in water, it percolates to those situations whence thermal springs derive their heat.

If atmospheric air, contained in water, descend to situations beneath the earth's surface where the whole is heated, the oxygen of that air may be separated, and nitrogen evolved. Supposing the water with the contained air to come into contact with such substances as sodium and potassium, we should expect to have at least a decomposition of part of the water, and a consequent evolution of hydrogen. Now as hydrogen, either free or combined with substances other than oxygen, and then constituting the water itself, is not found in all thermal springs, though it is present in some, we are led to conclude that the union of oxygen with a metallic base causing great heat, is not necessary to the existence of such heated waters. The volcanic conditions, above adverted to, no doubt may, and probably do, cause numerous thermal springs. The condensation of steam impregnated with the gaseous and vapourized matter thrown off by volcanos, would afford results so like the chemical contents of certain thermal springs, more particularly when we consider the effects which may be produced by the percolation of waters, so impregnated, through many rocks, that there is much difficulty in resisting the conclusion that some at least are due to similar causes.

To explain the origin of many thermal springs, even when nitrogen is present in them, and when sulphuretted hydrogen and certain like compounds are absent, other conditions would suffice. The percolation of water, impregnated with atmospheric air, to depths where it would take a high temperature from the effects of central heat, and the more quiet oxidation of substances by means of the oxygen of the contained air, would afford us effects adequate to explain those we observe in many thermal springs. In either case the elastic force of steam appears necessary for the propulsion of the water upwards through the cracks and fissures of the earth's crust, whether in mountain ranges or elsewhere. The theory of central heat would give us the necessary temperature at reasonable depths, and the decomposition of water by certain metallic bases would produce the same effect. We seem to make a fairer approximation towards an explanation of the phænomena of thermal springs when we combine central heat with the decomposition of water by certain metallic bases, such water being variously impregnated with foreign matter. The equal temperature of hot springs would be explained by central heat; for though thermal waters may be only condensed steam, still the power of the aqueous vapour to rise to a constant height in the fissure, through which the hot water escapes, must depend upon the regularity of the heat beneath. This effect could scarcely be produced by the mere oxidation of a metallic base, because the less the thickness of the resulting oxide, the greater the heat, from the comparative facility of combination; and consequently, after a time, the heat thus caused would decrease, at least until a fresh exposure of the metallic base be obtained, by fracture of the oxidized crust, or otherwise.

In volcanos we not only require great heat, but also chemical combinations and changes, which shall produce the gaseous and liquid substances evolved and thrown out when they are in activity. Aqueous vapour is in general very abundant. The principal gases hitherto detected consist, according to Dr. Daubeny, of muriatic acid gas, sulphur combined with oxygen or hydrogen, carbonic acid gas, and nitrogen. The sublimations of Vesuvius are, according to Sir Humphry Davy, common salt, chloride of iron, sulphate of soda, muriate and sulphate of potash, and a small quantity of oxide of copper. M. Boussingault, who has examined the gases and vapours discharged from Tolima, Puracé, Pasto, Tuquères, and Cumbal, volcanos of equinoctial America, considers that such gases and vapours consist of a great volume of aqueous vapour, carbonic acid, sulphuretted hydrogen, and sometimes the vapour of sulphur. He remarks that the sulphurous acid and nitrogen found in these craters are accidentally present; and that muriatic acid gas, pure hydrogen, and nitrogen, were not evolved from these volcanos*. The gases evolved from volcanos, and the sublimations discovered in

* Annales de Chimie et de Physique, tom. lii. p. 23. 1833.

and around them, have been examined by few competent chemists; and those to whom we are indebted for our information on this head have, in a great measure, confined their attention to a few volcanic vents of Italy, Sicily, other Mediterranean islands, and of equinoctial America. The field of observation has therefore been limited, more particularly when we consider the numerous volcanos scattered over the world.

It will be obvious that many other products than those above enumerated may be detected in volcanos. Indeed we can scarcely suppose that a variety of substances thrown together, and exposed to great heat, can do otherwise than afford products varying according to the conditions under which the volcano itself exists. However similar the general products of volcanos may be, many must be influenced by such circumstances. Fragments of rock broken from the sides of the vent are often thrown out with little change, the rapidity of their discharge having prevented any action of the heat or volcanic substances upon them; but it must very frequently happen that they fall into the incandescent mass, and thus assist in the formation of new compounds. According, therefore, to the nature and abundance of the disrupted rocks would be the resulting products. We might hence infer, what is matter of fact, that volcanic countries offer a rich field for the labours of the mineralogist.

When we observe liquid incandescent lava in a volcanic crater, we generally find that it is greatly agitated by the passage of vapour or gas through it. At intervals the volume of vapour or gas is greater than at others, and is discharged with a proportionate force, sometimes so great as to throw up a mass of the liquid incandescent rock, and to scatter it on all sides. The column of lava often appears as if in a great measure sustained by compressed gaseous matter beneath, which, when it has acquired sufficient force, uplifts the viscous substance above and escapes. The pressure of the superincumbent lava on the gases and vapours must often be very considerable; and there must always be a constant struggle between the compressing force of the lava and the expansive power of the heat, producing effects dependent on the relative intensity of each. We have seen that many gases are rendered liquid by a comparatively small amount of pressure; so that if by any chance, such as the dormant state of a volcano, some of them should be liquefied, their return to a gaseous state, produced by the application of the necessary heat, would cause explosions not only sudden but of extreme violence*.

The abundance of aqueous vapour discharged, and the common occurrence of volcanos in the sea or around coasts, naturally suggest the idea of the percolation of sea-water to certain metallic bases of the alkalies and earths. It has been supposed that the salts sublimed, and the muriatic acid gas evolved, give great countenance to this hypothesis; but the perco-

* Mr. Poulett Scrope has remarked (Considerations on Volcanos, p. 82.), that the great projectile force of the elastic vapour discharged from a volcano is occasioned by circumstances analogous to those produced by the ignition of gunpowder in a cannon, as far as regards the direction and force communicated to the discharged vapours or gases, the sides of the volcanic vent or pipe corresponding to those of the cannon.

lation of sea-water to incandescent matter, such as lava, would, under certain circumstances, produce somewhat similar results. When we obtain hydrogen in abundance, we seem to require the decomposition of at least a portion of the water. How far the sulphur requisite for the production of the sulphuretted hydrogen, often observed, may be derived immediately from the seat of volcanic action, or from the sulphate of soda contained in sea-water, is an interesting question, because it leads to an estimate, under this theory, of the substances situated beneath, and in the foci of, volcanic action, which it may be necessary to add to the contents of the sea-water itself, to account for the vapours and gaseous matter evolved.

We now know that volcanos exist at considerable distances from the sea. There is, according to MM. Klaproth, Abel Rémusat and Humboldt, a volcanic region in central Asia, between 300 and 400 leagues from the ocean, with an area of about 2500 square geographical miles. In the central chain of the Andes, also, the Peak of Tolima has been in a state of eruption more than once since historical times. Now it would be highly interesting to study how far the gaseous matters evolved from these vents, so distant from the sea, differ from those discharged from vents either surrounded by, or not far distant from, the waters of the ocean. Where saline lakes or inland seas are present, as might be the case in some parts of Asia, the results would be equivocal. The percolation of water through rocks impregnated with saline matter would also render the observations uncertain; yet it would be most valuable to have observations, by experienced chemists, on the products of volcanos differently situated as regards their proximity to the sea. By such means we should more surely approximate towards a knowledge of the causes of volcanic action.

The most marked substance which we should not expect the sea, from its known contents, to carry to volcanic foci, is carbonic acid, a gas often evolved in considerable abundance, particularly at the termination of eruptions, as if escaping after the great mass of matter has been thrown out. Now this is a substance brought freely to the surface of the earth both by thermal and cold springs. Its presence in the former is interesting as connected with volcanos; indeed the cold condition of the other waters containing it does not prove that such waters may not originally have been heated, and have become cooled in their passage to the surface. The weight of a column of water above, and the pressure of steam beneath, might force the water of springs so circumstanced to take up more carbonic acid than it would otherwise do, upon the same principle that artificial carbonated waters are prepared under pressure. The warmer the water, all other things being the same, the less would necessarily be the quantity of carbonic gas; and we see that the Carlsbad waters, with a temperature of 165° Fahr., contain five cubic inches of this gas in a wine pint*, while those of Pyrmont are impregnated with it to the amount of twenty-six cubic inches in the same measure, the spring not being thermal†.

One of the great characteristics of volcanic action is the violent evolution of gaseous matter and aqueous

* Berzelius. † Bergmann.

vapour. Possibly some of the latter may be produced by an union of free hydrogen with the oxygen of the atmosphere, and there would generally be no want of electric discharges to produce the combination, for such are the very common accompaniments of eruptions; but the great mass of it really seems water converted into steam, and driven off with no ordinary force. We should obtain aqueous vapour in abundance by the discharge of water on incandescent matter; but the production of some of the gases evolved by the mere decomposition of the salts dissolved in sea-water appears to be more difficult by the same means. Chloride of sodium is common as a sublimation in volcanos, and chloride of iron has been detected among the Vesuvian products by Sir Humphry Davy. The chloride of sodium is, however, nothing but dry sea-salt, so that we obtain it by the mere action of heat, and the evaporation of the sea-water. Muriatic acid gas is frequently observed in volcanic eruptions, but no estimate has yet been made as to its relative importance, and it would remain a question how far it might be produced in the volcano itself by a combination of hydrogen with chlorine, or be directly obtained from sea-water.

The relative amount of hydrogen gas evolved, either combined with sulphur, as sulphuretted hydrogen, or free, is important, because from it some estimate might be formed of the necessity, or supposed necessity, of the decomposition of water to account for the phænomena observed. If volcanic action depends on the decomposition of water, the oxygen uniting with the metallic bases of certain earths and alkalies, an enormous volume of hydrogen, viewed collectively, would be evolved. Dr. Daubeny has suggested that the hydrogen thus thrown into the atmosphere would combine with oxygen, form water, and be again conveyed to metallic bases, again to be liberated; so that a given amount of hydrogen would merely act as a carrier, and there would be no great accumulation of this substance in the atmosphere. One curious result would, however, be produced if oxygen be separated from the atmosphere by these or any other similar means: the volume of oxygen in it must decrease in a ratio proportionate to the number and intensity of volcanic eruptions.

Information respecting volcanic vapours, gases and sublimations is undoubtedly very limited, and not exactly such as we require for a full consideration of the subject: but as far as it extends, the percolation of water to volcanic foci appears requisite for the phænomena observed. Whether we are to suppose the water decomposed, at least a portion of it, thus producing heat, or whether it merely falls on incandescent matter, is a separate question. Elastic vapours or gases seem necessary to the propulsion of the solid and liquid matters ejected. Their discharge always accompanies a great eruption, and their propulsion of liquid or solid matter into the atmosphere is one of the most marked and imposing features of volcanic eruptions. Now the mere fracture of the earth's crust, even to liquid incandescent matter does not appear sufficient in itself to produce the effects we witness in active volcanos, unless we suppose the elastic vapours and gases, so commonly discharged, to be constantly present beneath the earth's solid crust, over all parts of the globe. This supposition implies a state of things very difficult to conceive, more particularly when we consider the kind of vapours and gases evolved. At all events, it does not offer the same simplicity as the hypothesis of the percolation of sea or other waters, through fractured strata, to volcanic foci, where they undergo changes, whatever their immediate cause may be, producing at least a large proportion of the elastic vapours and gases so important in all volcanic eruptions.

We may here remark that sub-aqueous would so far differ from sub-aërial volcanos, that the vapours and gases evolved from the vent would, for the most part, be readily absorbed in the first case, and not escape into the atmosphere, unless the crater be so near the surface of the sea, that the force with which the vapours and gases are expelled is sufficient to overcome the resistance of the superincumbent weight. At moderate depths this resistance would however be sufficient, more particularly when added to the facility with which most of the vapours and gases would combine with water. Aqueous vapour would readily disappear, and only be carried high above the discharging crater when near the surface of the water, as was the case during the formation of the island of Sciacca, in 1831, between Sicily and Pantellaria. Such aqueous vapour condensed is in the condition of water from which air has been driven by ebullition; it is in fact distilled water. Its temperature would soon be reduced by the surrounding mass of sea, and it would speedily be, though somewhat mixed with the latter, in a condition to absorb muriatic acid, sulphurous acid, sulphuretted hydrogen, and carbonic acid, independently of any absorption that might take place by the sea-water itself. Free nitrogen would only be sparingly absorbed, but if it be united with hydrogen in the form of ammoniacal gas, it would speedily be so. In deep waters there would necessarily be great pressure, and this alone would tend to force the gaseous products into the water. Indeed, as far as regards mere pressure, sulphurous acid gas would be liquid beneath 68 feet, and sulphuretted hydrogen beneath 578 feet of water, at the respective temperatures of 45° and 50°; temperatures common in deep waters, and to which we may conceive those around an active volcanic vent would be reduced at no consisiderable distance from it. It is almost useless to notice the effect of pressure by water of such substances as muriatic acid gas, or ammoniacal gas, supposing the latter to be formed, for they would rapidly disappear by absorption in it; but they would both become liquid beneath depths by no means considerable, the former at 1360 feet, the latter at only 221 feet.

Viewing the subject solely with regard to pressure, there would be more difficulty with carbonic acid, for it required a temperature of 32° and a pressure equal to 36 atmospheres to render it liquid in Mr. Faraday's experiments. The pressure would be readily obtained beneath a moderate depth of sea, but such low temperatures would only be found under certain conditions in polar latitudes. Considering, however, that great additional pressure would produce the same effect in forcing the particles of carbonic acid together as the lesser pressure and greater cold above noticed, we may infer that beneath 500 or 600 fathoms of water this gaseous substance would be liquefied. Now this is no extraordinary depth; on the contrary, it must be very common over a considerable portion of the area occupied by sea. Even in the Mediterranean deeper water is not rare. M. Berard found no bottom with a line of 1200 fathoms in situations between the Balearic Isles and Algiers, when sounding for the purpose of ascertaining the temperature at different depths. As nitrogen has never yet been liquefied by experiment, we do not know that it would become so beneath the water of any part of the ocean; but though it may not become liquid beneath such pressure, it would, being elastic in its gaseous condition, be rendered so dense as to be subject to those effects which have been attributed to atmospheric air under similar circumstances.

The evolution of vapours and gases, and the causes which may have produced them, are, however, not the only objects of interest in this inquiry. The probable production of the solid and liquid matter ejected is also highly important. Streams of lava are rarely if ever produced, except at those periods of intense volcanic action termed eruptions, and it is very generally considered that these masses of liquid rock are derived from considerable depths beneath the surface of the earth. Prior to great eruptions, the volcanic vent is generally clogged, as it were, by solid lava and fallen rubbish, so that there is both solidity and weight opposed to the free passage upwards of gases, vapours, or liquid melted rock. We can scarcely consider the crack, or vent, to be void of liquid or solid matter from the orifice which is so choked up to the depths whence we suppose the lava, vapours and gases to be originally derived. We should rather infer that the pipe or crack was filled with lava intermingled with vapours or gases, which may, or may not, be able to pierce through the matter accumulated in the crater.

That lava retains its heat for considerable periods of time, we have abundant evidence, even when it is ejected from the crater or sides, and can thus radiate its heat freely into the atmosphere*. If it can retain its elevated temperature when thus exposed, what length of time may we not allow for its doing so within the pipe of the volcano itself, surrounded on all sides by matter greatly heated, and like itself, an exceedingly bad conductor of heat. Even in those cases when centuries elapse between the great eruptions of any given volcano, the lava is probably liquid beneath no very considerable depth. The particles of the higher portions may be, and probably are, so firmly united, even when red hot, as to require very great force for their separation, and this perhaps may be one of the reasons why explosions in the earlier stages of an eruption are so violent, powerful resistances being suddenly overcome not only by the strong pressure of the liquid lava beneath, but also by elastic vapours struggling to free themselves.

If the cone or elevated portion of a volcano bears, as is commonly supposed, but a small proportion, in

* Mr. Poulett Scrope observed a current of lava, flowing at the rate of a yard a day, on the flanks of Ætna, in 1819, after it had been ejected nine months. (Considerations on Volcanos, p. 101.) Dolomieu and Ferrara notice other currents of lava, also ejected by Ætna, as in movement ten years after they were thrown out. Sir William Hamilton lighted small pieces of wood in the fissures of a current of Vesuvian lava four years after it had been ejected.

height, to the depth whence the lava is derived, the column of liquid rock would in all cases be considerable, though its height would vary materially in different volcanos, depending on the elevating power beneath. Now if the crust of the globe were of the same general thickness throughout, and the interior of the earth fluid, any cause acting uniformly in the latter, so that it should endeavour to pierce the former through fissures, would, all other circumstances being equal, maintain columns of similar fluid matter at equal altitudes; so that we should expect volcanos to be of nearly equal altitudes on the surface of the earth, if they were produced solely by the endeavours of an interior incandescent fluid to escape outwards through fissures or cracks on the surface above. Such a state of things supposes expansive power in the interior mass, and consequently a cause by which its volume is increased. This cause could not be radiation of heat from the earth, which would tend to produce the contrary effect; it must be something causing a comparatively wide separation of the particles of matter in at least the external portion of the fluid interior mass.

If the solid surface split while cooling into numerous fragments, such fragments might be supposed to sink or swim in the liquid matter beneath, according to its relative specific gravity. Taking the mean density of the earth's mineral crust at 2·6, fragments of this specific gravity would sink in felspathic or trachytic lava and float in basaltic matter. Considering however, that lava, whether trachytic or basaltic, would be rendered so much lighter by heat, (as they must always be to a certain extent,) that masses of cool matter of the density of 2·6 would sink in them, these masses would all tend to come together from the action of gravity; and as this force must always exist, one does not see that the fragments could be separated at all, unless an expansive force acting from beneath compels them to do so; for they would be squeezed together.

An expansive force of this kind, acting generally beneath the crust of the earth, the latter, broken and fractured as it is, could scarcely be expected to resist the power exerted upon it in every direction. It must, we should conceive, give way before so general an application of force, liquid matter being introduced among the fissures in all directions, and the solid superficial portions separated to enable it to do so; the result being a state of comparative rest, produced by the expansion having attained its limits. This theory seems directly opposed to that which considers the present condition of the world to arise from the radiation of heat, and a consequent diminution in the volume of the earth; for according to it the volume would be increased, producing all the consequences of such increase. And should we suppose similar effects to recur at intervals, the volume of the earth would become gradually larger. Perhaps it may be considered that the percolation of water to the metallic bases of the earths and alkalies, situated generally beneath the oxidized crust, would cause expansions both from the heat evolved and the combination of the metals with oxygen; the unequal checks offered to the passage of the water producing effects more resembling those we witness. Radiation of the heat so caused would certainly produce contraction, and thus we might obtain a plausible account of some of those alternate rises and depressions of continents and islands, which geology teaches us have taken place on the surface of our planet; for while intense heat was produced by the combination of the oxygen of one charge of water with the metallic base, no more water could approach the lower body from above, until the heat was sufficiently radiated or conducted away, and therefore there would be no gradual and continued expansion unchecked by contraction. It would also serve to explain the unequal rise and depression of masses of land. It would still, however, give us an increase of solid matter at the expense of the atmosphere, in the manner previously noticed (p. 153), continuing to rob it of its oxygen to an enormous amount. It is apparently a great objection to this view that we should scarcely obtain the great inequalities we now observe on the surface of the earth. We have already stated the difficulty of accounting for mountain ranges and remarkable lines of elevated strata by these means: there is also a difficulty as to the bed of the ocean; for water, all other things being the same, should percolate easier to the metallic bases beneath deep seas, even supposing the oxidized crust to be there of equal thickness, than through continents and islands.

Should this cause produce the elevation of great masses of land, we should expect the latter to exhibit the marks of volcanic action at the same time; that is, we should expect the hydrogen evolved to escape through fissures, accompanied by fused rock and various gases and vapours. Such an hypothesis might plausibly explain the elevations of parts of Italy, but appears inadequate to account for the remarkable rise of land, within historical times, noticed in Norway and Sweden*, it being unaccompanied by the evolution or ejection of gases or other substances, which might lead us to believe that decomposition of water beneath, from the causes above noticed, produced heat, and that heat the necessary expansion.

That changes in the relative positions and levels of sea and land have taken place on the surface of our planet, all our continents, and the greater part of our islands, bear abundant evidence. By far the larger portion of dry land must have been formed beneath the waters of the ocean; not only do abundant marine organic remains, but the structure of the rocks themselves, prove this: indeed the comparative area occupied by rocks which could have been formed at those levels above the ocean which they now occupy, is exceedingly small. They would necessarily consist of volcanic products, of deposits in lakes and masses of fresh water, which have disappeared and left them exposed, and a few incrustations formed by the evaporation of water charged with foreign substances, such as carbonate of lime. The rocks which have been clearly produced beneath the sea, by chemical or mechanical deposits from it, are stratified; and when we compare these with the stratified rocks which may have been formed out of it, the difference in the volume of the two products is enormous.

We have elsewhere observed that an exceedingly

* This elevation is considered to be unequal, being estimated at the rate of 4 feet in a century in tne northern part of the Gulf of Bothnia, thence diminishing towards the south, so that it is only 2 feet in the same time on the coast of Kalmar, and not observable at the islands of Œland and Gottland.

small and unequal expansion on the surface of the earth might produce great continents*; and if the reader will turn back to fig. 21, he will perceive how readily this may be accomplished. The boundary line of the circle represents a proportional depth of one hundred miles. Now it will be evident that if any circumstance should cause an increased thickness of $\frac{1}{50}$ in this depth so that the exterior be raised $\frac{1}{50}$, we should have an elevation produced to the height of two miles, sufficient to raise a great body of matter deposited beneath many parts of the sea above its level. This thickness, however great it may appear to us, accustomed as we are to measure elevations even by fractions of our own mean height, is in fact so small, relatively to the diameter of the earth, that the unavoidable roughness in the printed line of the circle (fig. 21.), represents inequalities of far greater amount.

The unequal contraction of the crust of the globe, (really of very trifling importance compared with the volume of the earth,) would not only appear to raise large areas, composing continents, bodily out of the water, by producing great depressions, but would squeeze the principal surface fractures into mountain ranges. The general movement might be so slow that we should have difficulty in measuring it by our ideas of time, while here and there the pressure of the sides of the great fissures against each other would cause more sudden movements, the force applied, however general its nature, overcoming the resistances opposed to it at unequal intervals; thus producing those sudden dislocations which, measuring

* Sections and Views illustrative of Geological Phænomena, p. 71.

them by his own height, or more commonly some minor portion of it, man terms enormous. Taken by itself, this cause would act on too large a scale to produce volcanos: these seem rather the effect than the cause of great movements in the crust of the globe, the conditions for their existence being apparently produced by the effects resulting from a much more general and greater exertion of power. While, however, the more important effects may be attributed to the greater, smaller disrupting effects may be assigned to the minor cause; and probably no small amount of the elevations, depressions and fractures observable on the surface of the earth may be due to the compound action of both; not forgetting that the dislocations and movements of superincumbent rocks may be such as readily to permit the intrusion or injection of igneous matter of different kinds among them*.

* To prevent misconception, it may be as well to state that the foregoing remarks were in type, precisely as they now stand, before Mr. Babbage (as I had occasion to make known to that gentleman at the time of reading his paper before the Geological Society,) communicated his very interesting views to me respecting the elevation and depression of land from variations in the expansion and contraction of rocks; variations which depend on the different conditions under which the rocks have been placed beneath surfaces possessing different powers of radiating heat. The passage alluded to in the text, and to be found in the work quoted, is as follows: "To one who looks at such a diagram, it will be obvious that slight and unequal contractions of the mass of the earth would produce changes we should consider important; and it may occur to him that mere thermometrical differences beneath the earth's crust might be sufficient to raise whole continents above the level of the sea, or plunge them beneath it." Still further to prevent misconception, it may be as well to state that I have every reason to believe Mr. Babbage never saw the passage until I pointed it out to him, *after* he communicated his views to me; and that I do not pretend to have entertained opinions similar to his respecting the probable effects of the causes he notices *before* he stated them to me.

CHAPTER VIII.

When we regard any good representation of the real physical features of a mountain-chain, such, for instance, as a correct map of the Alps, we are struck with the great resemblance which the principal valleys bear to fissures. There are commonly some valleys which hold the same course with that of the main range of mountains, while others run in directions nearly at right angles to it. These lines will often be found not strictly straight when examined in detail; but when viewed on the great scale, the tendency of the leading valleys to take longitudinal or transverse directions is very remarkable; so much so, that physical geographers have employed the terms 'longitudinal' and 'transverse valleys' without reference to their origin. When we attentively study the cause of these phænomena, we are generally able to detect lines of dislocation or violent contortion coinciding with those of the valleys. The lines of the principal valleys are therefore in such cases the lines of principal fracture; in fact they are precisely such as we should expect to find resulting from those causes of mountain-chains which have been previously noticed.

The great fractures generally run, as might be expected, in the lines of longitudinal valleys; that is, the fissures productive of the greatest amount of vertical movement, throwing the planes of once continuous strata either up or down, as the case may be, are in such directions. The transverse valleys, unless the mountain-chain has been traversed by systems of cross fractures of various geological epochs, are not in general so remarkable for vertical as for horizontal movements. The beds on either side more closely correspond, and there is more resemblance to a secondary fissure; that is, a fissure produced by the rending of the masses split longitudinally.

When we attempt in imagination to replace the dislocated rocks composing mountain-chains, so that the edges of the disrupted strata should be brought into contact and again form continuous planes, we are soon sensible that a large amount of matter has been removed, and that we could not reproduce one great continuous mass of the rocks composing the mountain range.

Fig. 28.

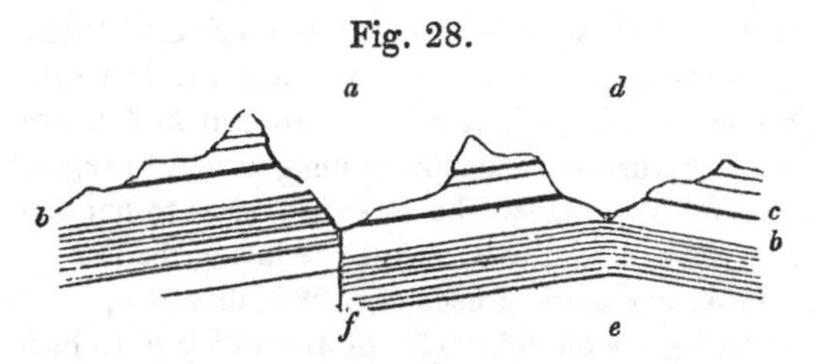

Let *a* (fig. 28.) represent a mountain valley, in which a fault, or dislocation, *f*, runs along the bottom, and that in consequence of this dislocation a well-known bed of rock *b* occurs at different levels on either side. It is clear that if we attempt to bring the bed *b* on the one side in a line with *b* on the other, either by lifting up one mass, or letting down the other, there would be no continuity of the general mass of beds; a great portion of matter would evidently be wanting.

Fig. 29.

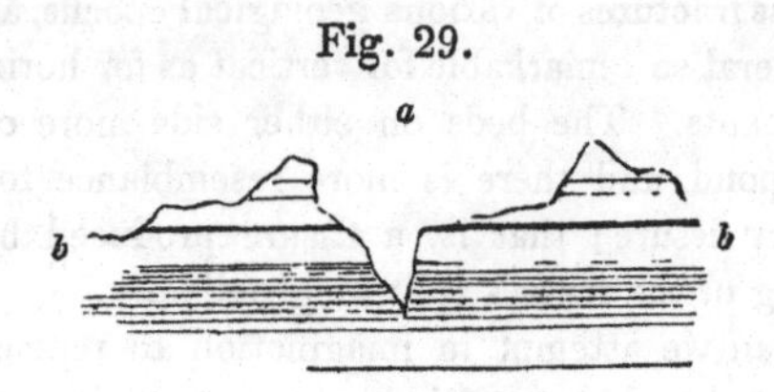

The annexed diagram (fig. 29.) represents the bed *b* placed in the same plane in which it existed prior to the fracture, and the space *a* will convey an idea of the amount of matter removed not only beneath the plane of *b b*, but also above it. If, instead of a fracture producing the sudden rise or fall of great masses of rock on either side, we consider the opening caused by a great bend of stratified matter, fracturing and separating the upper parts from each other, (a fact often observable in mountain-chains,) we still find matter removed when we attempt in imagination to replace the bend and reduce the stratified mass to horizontality, as will readily be seen by reducing the lines *b c* representing marked beds (fig. 28.), driven upwards by the bend *e* on either side of the valley *d*, to horizontality.

Now if we form an approximative estimate of the cubic contents of the matter contained in a mountain-chain, such, for instance, as that of the Alps, above any given level, and roughly calculate the amount of matter that might be contained in the mass, if it were merely broken to pieces, the dislocated parts fitted to each other, and forming as once a continuous whole, we find the portion carried away or missing very considerable. Such a calculation is necessarily very rough from the nature of its elements; the amount of contortion is more particularly embarrassing; yet sufficient may be accomplished to show satisfactorily that there has been a very large amount of matter removed.

Moving water is the only agent known to us capable of carrying away this great collective mass of rock. In order, therefore, to form a just conception of the time and conditions required to produce the effects observed, we should carefully examine the latter, and estimate the transporting powers of those running waters which now exist among the mountains themselves, and which transport detrital matter from the central parts outwards.

Beneath the envelopes of snow, commonly termed perpetual, because a given amount of them never becomes liquefied under the ordinary circumstances in which they are placed, the surface of solid rock is greatly protected, if not from decomposition, at least from removal, except when an avalanche may cut deeper than usual, or a mass of the rock supporting the snows fall into the valley below. So long as ordinary circumstances continue, the dome of Mont Blanc will be covered by snows, which began to accumulate there after the mean temperature of the climate and the altitude of the mountain were the same as at present. Such snows would protect the supporting rock from removal if they had not a tendency to descend in the shape of glaciers, and thus, by friction on many points of the surface beneath, force portions of such surface to a lower level. Viewing the subject in this light, we can scarcely consider any considerable area of rock beneath perpetual snow to be absolutely free from degradation, though such conditions would be generally unfavourable, and the amount of detritus so produced be relatively unimportant.

Lofty peaks, or aiguilles, on which snow cannot rest, merely insinuating itself among the vertical or highly inclined crevices, and a zone comprised between the line of a constant snowy covering, and that, at a lower level, where the rapid alternations of frost and thaw terminate, are the parts of a high mountainous region generally most productive of large detached fragments, all other circumstances being equal. If the sides of a mountain be precipitous and exposed to the latter condition, the collective amount of fallen matter will be found very considerable at those points where it has, at least for the time, been brought to rest. During those seasons of the year when the melted snows above cause a descent of water to lower levels, as well over the general surface as among the various drainage channels, the water, insinuating itself into the interstices of the surface rock during the day, is frequently frozen at night. Water, entering into its solid state, expands, as is well known; therefore all those portions of rock which thus become heaved out of their original places by the formation of ice during the night, will, if their centres of gravity permit, fall when the heat of the succeeding day liquefies the ice. Should the centres of gravity of the rock-fragments be so situated as not to produce an immediate fall, the repetition of the above effects would tend not only to loosen but also to force them into positions where, if they do not fall by their own weight, they may do so by the action of powerful winds, by a blow from a fragment detached from above, or by the loss of part of the support beneath.

Where the temperature and other circumstances permit the growth of plants, the fall of rocks, from the cause noticed above, is checked. The roots, it is true, act most powerfully in dividing fragments from each other, but at the same time they bind masses of them together, not readily removed except under extraordinary circumstances, such as violent storms and the like. The less the slopes are inclined, the greater is the stability of the vegetation upon them; and, consequently, the greater the protection afforded to the rocks beneath from this cause, the less the amount of rock removed to lower levels. Broken bare rocks, situated between the lines of perpetual snow and of a fair protecting amount of vegetation, other things being equal, will be the exposed area among lofty mountains where we should expect to find the greatest amount of degradation. The size of the masses detached will necessarily depend upon the precipitous character of the mountains, the relative elevation of the peaks and cliffs, the nature of the rock of which they are composed, and the kind of climate in which the particular mountains are situated.

Atmospheric causes tend, therefore, constantly to destroy and throw down to lower levels such parts of rock-masses as are by any means thrust up sufficiently high into the air, more particularly when there is a zone in which there may be almost diurnal changes from frost to thaw. In point of fact, all parts of a mountain range have a tendency to be degraded more or less by the action of atmospheric causes upon them; and even the portions protected by vegetation may be eventually undermined, so that the fragments of rock bound together by the roots of trees are moved in a mass, should the trees be uprooted and washed onwards by descending waters. It should also be recollected that in mountainous regions the slopes, even those supporting vegetation, are for the most part highly inclined, so that mineral and vegetable matter, finely divided, is readily moved downwards mechanically suspended in myriads of little threads of water, which pass almost imperceptibly among the vegetation itself.

We have hitherto merely noticed the mechanical action of atmospheric causes; the chemical changes are, however, no less important. We have seen (p. 92.) that by the percolation of water numerous rocks are continually suffering an alteration in their component parts. Many of these changes produce ready disintegration, so that substances, apparently of great toughness and durability, give way before moving water, which otherwise would long resist its mechanical action. The quantity of oxygen contained in the air, which rain-waters have absorbed from the atmosphere in their fall, will greatly assist in forming chemical combinations with certain constituent portions of rocks, which thus become more readily decomposed than would otherwise happen, as has been remarked by Dr. Boase*. The carbonic acid, also, absorbed by rain-water from the atmosphere must likewise assist in the decomposition of many

* Contributions towards the Geology of Cornwall.

rocks; and as snow-water appears richer in oxygen, thus derived, than common rain-water*, we should expect all such decomposing effects as can be produced by oxygen so circumstanced to be greatest in rocks where snow-water percolates, and consequently in those elevated portions of mountain ranges which become saturated with melted snow.

We thus see, that if a mass of fractured rocks be thrust up into the atmosphere, they will immediately be acted on by the latter; and such causes of destruction as we now witness would, if time be allowed, be sufficient to crumble the exposed portions into fragments of various sizes, removable either by gravity, moving water, or the union of both causes, to lower levels. The more sudden the relative elevation of the mountain range, the greater the collective area of the fresh-fractured parts exposed to the effects of the atmosphere; assuming that the mass has not been raised upwards bodily without cracks and fissures, but that disrupting forces have acted on it in various directions. In such a state of things, and before vegetation could in any way protect the newly broken parts, the degradation from meteoric influences, as they have been termed, would be considerable, the broken masses and smaller fragments would be lodged wherever slopes or levels were favourable, and, consequently, the greater amount would be eventually carried to the bottom of the open part of the dislocations, and constitute the bottoms of valleys.

That a system of cracks or fissures, produced in

* MM. Gay-Lussac and Humboldt found 34·8 per cent. of oxygen in the air contained in snow, and 32 per cent. of the same substance in that discovered in rain-water.

the manner above noticed, could afford so complete a drainage that bodies of accumulated still water should not exist in the various channels, can scarcely be supposed. Lakes of greater or less magnitude must be formed in numerous valleys, and their bottoms be the receptacles of the detrital rubbish swept into them by the waters on their onward course to lower levels. Such lakes, either by cutting through their barriers, or from being slowly filled by transported detritus, would gradually disappear. In the first case we should trace the evidences of their former existence by various marks on the sides of the mountain valley, or by detrital matter cut through; in the second we should have a plain from which the mountains rose suddenly on either side, the river passing through it. There would probably also be another cause of their disappearance, productive of effects having considerable influence on the condition of lands at lower levels. A sudden removal of a lake's lower barrier by disrupting forces acting in the line of the valley (precisely the line we should expect to be acted on by subterranean forces, being that of least resistance,) would not only cause the lake to disappear, but would also throw a body of water violently upon the lower levels. The effects produced would necessarily depend on the volume and velocity of the discharged water: both would of course vary according to circumstances; but we should anticipate that, under these assumed conditions, they would be sufficient to cause considerable changes, tending to force forwards the contents of lower lakes wholly filled up, conveying a great body of detrital matter to the lower levels, and transporting blocks which could not be moved by the power of the rivers, even during times of flood.

Fig. 30.

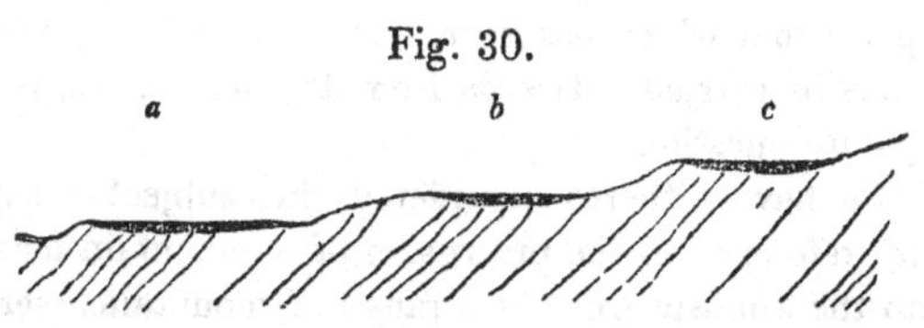

If *a*, *b*, *c*, (fig. 30.,) represent lakes produced by the inequalities in the bottom of a mountain valley, it is clear that they may be all contemporaneously (geologically speaking) filled by detrital matter, mixed with exuviæ of the terrestrial and fresh-water creatures inhabiting the neighbouring country or the lakes themselves at the same epoch. Unless, however, there be, as not unfrequently happens, abundant detrital matter carried laterally into the lakes from the heights on either side, the higher lakes would have a tendency to be sooner filled than those at lower levels, being the first to receive the transported matter brought down the valley.

The effects which we have thus supposed would be produced by such an action on great masses of rock, that they are fractured and thrust into the atmosphere, constituting ranges of mountains, bear so close a resemblance to many facts we observe in the Alps and other considerable ranges, that we are irresistibly led to consider them the result of somewhat analogous causes. Hence a great body of matter might be detached from mountains, more particularly when fractures were fresh, and transported outwards by a long-continued action of those causes now constantly in operation on the surface of the globe, and more particularly observable in high mountain ranges. Whether the total amount of solid matter necessary to replace that which has been removed, could by such means be carried outwards from the mountains, is a separate question.

We have hitherto considered this subject solely with reference to the protrusion of a mountain mass into the atmosphere. If a range of mountains were to be now formed, the chances, considering the great relative proportion of superficial water to land on the earth's surface, would be against its protrusion, at least in the first instance, into the atmosphere. Now if a long line of longitudinal fractures, accompanied by transverse dislocations, be produced beneath the waters of the ocean, such waters would necessarily be agitated with more or less violence in proportion to the sudden elevation, volume and height of the upheaved and dislocated mass. All loose fragments would be washed to the lower levels, the bottoms of the valleys would be filled with detrital rubbish, and much might be transported to the lower levels on either side of the sub-aqueous mountain range. The denser the water, and the more impregnated with matter mechanically suspended in it, the easier would the larger fragments be carried by resulting currents, the differences of specific gravities being less. There would be great inequalities of surface; but, if depths be sufficient, the mechanically suspended matter would gradually settle, the finer portions uppermost.

If a range of mountains thus produced be brought gradually, or by successive movements of elevation, into the atmosphere, we should first have the destructive action of superficial tides, currents and waves upon it, and no small amount of matter would thus be removed, particularly the unconsolidated portions. If brought more suddenly into the atmosphere, the destructive effects would be greater; but in both cases, should the range once constitute a mountain mass rising above the sea, meteoric influences would commence, and be followed by the necessary consequences. Inequalities in the great valleys would produce lakes, detrital matter would be carried into them, the sides of the mountains would suffer degradation, and altogether, after a sufficient lapse of time, the effects produced would so closely resemble those previously noticed, that there would be great difficulty in deciding whether the original dislocations were sub-aërial or sub-aqueous. Should, however, rocks have been deposited in the latter case, accompanied by marine remains, decidedly in the great lines of valley, there would be little doubt that such parts at least once constituted a portion of a sub-aqueous mountain range, and continued in that state sufficiently long to permit the quiet deposit of rocks upon it.

We therefore see that, whether mountains be produced in the sea or on dry land, much solid matter must be borne away from it; but probably the greatest amount of detritus would be suddenly carried off by such a relative change in the levels of land and water, that the great dislocations should be partly sub-aërial, partly sub-aqueous; so that heavy destructive waves, of a magnitude proportionate to the disturbing force, should rush up the newly formed fissures, and sweep out a mass of fragments. In all cases, however, the running waters necessary to produce the numerous effects we observe in mountain-chains must have been due to atmospheric causes, the erosion being of that kind which marks the long-continued action of narrow streams of water moving with considerable velocity, and the transported matter, found among the mountains themselves, being arranged in the manner in which it would be if deposited by such streams. It might hence be inferred that we have a measure of the time during which any given mountain range has constituted dry land, estimating the length of time by the amount of detritus accumulated in any given position. A little consideration will, however, readily show us that the results are much too complicated to bear us out in other conclusions than that a large amount of detritus, accumulated by the comparatively slow action of mountain rivers, has required a long time for its production, without enabling us to estimate the amount of that time otherwise than very generally. The facility with which detritus could be produced in one situation would cause an accumulation of it, by a given force of water, to be far larger than another accumulation formed by deposition from an equal force of water, in equal times, where the production of detritus was far more difficult.

There being facts observable in many mountain-chains, and down great lines of valley, which would seem to require the force of a sudden rush of water, having considerable volume, for their explanation, recourse has often been had by those anxious to explain all geological phænomena by such intensities of force, and other circumstances, as we daily witness, to what has been called the 'bursting of lakes.' The idea seems to be that there is some very powerful force inherent in the body of the lake itself, capable of driving out the solid barriers composing its sides. If it had been recollected that almost all lakes are,

Fig. 31.

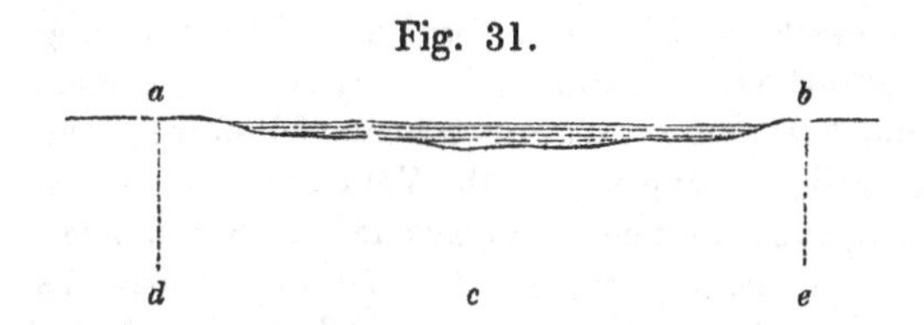

when viewed on the large scale, mere shallow basins of water, the sides of which rise at small angles, not unfrequently less than that represented in the annexed diagram (fig. 31.), it would readily be seen that the pressure of the contained water would be so small relatively to the resistance of the sides, that no fracture could possibly be produced by it. If, instead of the sides being, as represented above (fig. 31.), nearly horizontal, as is so commonly the case in nature when we make a section through the discharging outlet, we suppose them vertical, as at *a d* and *b e*, so that the lake be supported by the solid body *a d c e b*, we still find that the sides would be capable of resisting the pressure of the water against them, if the whole basin of rock be, as is commonly the case, sufficiently solid.

When we make proportional sections of all the great lakes known to us, we find that it would be impossible for the waters contained in them to break their barriers by any pressure they could exert on them. It could only be under very extraordinary circumstances that the rocky barriers of a lake could be forced outwards, as we have had occasion to notice elsewhere*. If common calculations had been made relatively to the amount of pressure which the waters of lakes could exert on their natural barriers, we should probably have heard less of the 'bursting of lakes' by such means. The sudden discharge of accumulations of water, caused by the fall of ice across a valley, as happened in the Val de Bagnes, seems, from want of attention, to have led to erroneous impressions respecting the power of water to drive its barriers before it. From the lake so produced, 530,000,000 cubic feet of water were discharged down a narrow mountain valley in about half an hour, and must of necessity, under such circumstances, have carried huge masses of rock before it. It rushed from the glacier with a velocity, according to M. Escher de la Linth, of 33 feet per second, so that when we take both the volume of water and its velocity into account, we have a transporting force of very great power; but in what manner such an event can illustrate, as it has been thought to do, the power of water to burst a rocky barrier, it is difficult to conceive. Lakes caused by the fall of masses of rocks from mountains across valleys may, to a great extent, be suddenly discharged, if the resistance of the dam thus produced be unequal to the pressure; the same effects would be produced if detritus, discharged from a lateral valley, dam up the main river, and the bar be unequal to the force exerted upon it; but these cases are very different from the sudden discharges of

* Geological Manual, p. 63.

such lakes as those of North America, or even those on each side of the Alps.

Although the conditions for the sudden discharge of lakes by the power of the contained water to force a passage through their lower barriers must be exceedingly rare, and the consequent sudden discharges themselves be equally so, lakes may be readily drained in a short time, as previously noticed, by fractures of their sides, caused by earthquakes or more intense exertions of similar forces. When we attentively examine mountain-chains, we find, at least all those as yet properly examined are found, to bear marks of more than one action of elevating forces upon them, raising them to the relative heights they at present occupy. We judge of the geological periods when such forces have acted by the kind of rocks quietly covering the disrupted edges of previously existing strata. Both rocks, or their equivalents in the geological series, being known, we are certain that the particular elevation took place after the consolidation of the inferior rock, and prior to the formation of the strata reposing on its broken and upturned edges; and we, moreover, have evidence that after the older rocks were upset, they remained in that state sufficiently long to permit the deposit of the strata resting on them.

Now finding strata resting on the disrupted edges of other beds on the flanks of a mountain-chain, only shows that some force upset the one before the formation of the other, and not that the first constituted a mountain range before the newer rock was deposited. Nothing is more common than to find two rocks in discordant stratification, as it has been termed, still constituting comparatively level tracts of country. If such a district be suddenly thrown up into a mountain range, inferior matter being introduced in the line of the main longitudinal fracture, or the edges of the same fracture merely squeezed upwards against each other, we should err in our inference if we considered such elevation to have been produced at two periods instead of one, merely because we found rocks in discordant stratification on the flanks. There ought to be clear evidence in the interior of the mountain range itself.

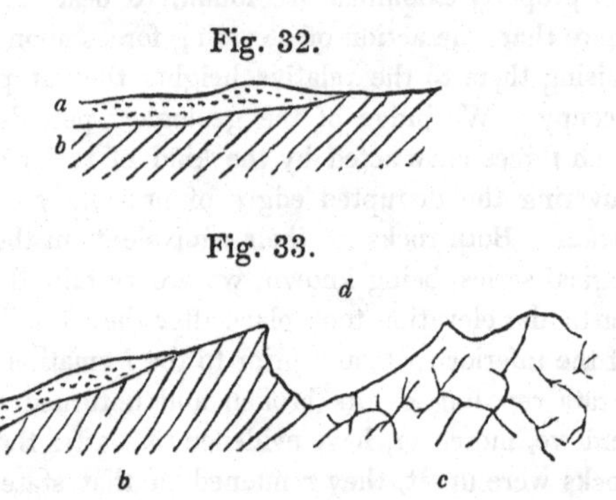

Let *a* (fig. 32.) represent a bed of rock resting unconformably on previously upturned strata *b*, the whole constituting the base of a comparatively level tract of country; and *a*, *b* (fig. 33.) be the same rocks tilted up by the protrusion a rock *c*; we should have appearances that might be considered the result of two elevating forces on the same system of mountains. They would in this case be the consequences of two disrupting forces, though only the last has produced the mountain range; but if the inclined strata (*b*) had been produced in any manner resembling that previously noticed (p. 50.), there would only have been one really disrupting and elevating force.

If mountains be, however, subjected to more than one movement of elevation, no matter how caused, and the disrupting force sufficiently intense, the separation of parts being easiest in the great lines of valley, any system of lakes formed at one period might be suddenly let out at another, and the volume of water discharged outwards carry with it a considerable amount of detritus, distributing it into a sea, or over dry land, as the case might be.

When we regard those immense masses of conglomerate so common along the flanks and skirts of mountain-chains, be their relative age what it may, we find them composed of fragments, either rolled or angular, derived from the rocks constituting the mountains themselves. According to the continuity and length of particular beds, should we feel inclined to judge of the character of the causes to which they are due. When continuous for considerable distances with common characters, we infer a more general force than when we see them interrupted at short and irregular intervals. The more distant a conglomerate bed may be from the mountains whence its materials have been derived, being at the same time similar to others nearer the range, the greater we should consider the intensity of force necessary to transport its component parts, all other things being equal. When sandstones and conglomerates alternate, we infer inequalities in the power of transport, and consequently that the causes of the power act irregularly. The volume, figure and specific gravities of the fragments would afford us some measure of the relative transporting powers if we were certain that the bottom over which the whole was borne, prior to deposition, was generally similar. Great irregularities in the volume and figure of detrital matter would necessarily be found in deposits at the mouths of torrents; and after a time, when levels of fair extent were produced, there would be a difficulty in transporting pebbles of large size outwards. Hence if we found continuous beds of conglomerate ranging in straight lines at moderate distances from mountains whence these materials have been derived, we may infer either that some force, more general than that possessed by the rivers descending from the chain, has produced them, or that the detrital fragments having been borne outwards by rivers from the mountains, some general force distributed them in a more regular manner than could be accomplished by the rivers themselves.

If detrital matter be brought down by numerous rivers from a mountain range into a sea, there would undoubtedly be an endeavour on the part of the waves of such a sea to distribute pebbles, even of large size, along shore, thus producing lines of pebbles parallel to the shore. And we might consider that if the supply of pebbles be abundant, the distribution of them would add to those previously thrown on shore, thus obtaining beds of conglomerate. To obtain regular stratified beds of sandstone by this process on the large scale would appear to be difficult, though interstratifications of sands and pebbles may sometimes be seen in the sections of beaches.

When we regard the various causes which may produce a transport of detritus from mountain chains outwards, it seems somewhat hasty to determine, *à priori*, that only one cause has to any extent ever been in operation. We ought rather carefully to study the various circumstances connected with deposits, evidently derived from the mountains which they skirt, —in fact, the effects produced,—and thence ascend to the probable causes of the facts observed, than to predetermine the causes without sufficiently examining the effects. The latter is too frequently the course adopted, and is the necessary consequence of seeing only through one medium. We may possibly by such means, and amid a thousand hypotheses, occasionally hit the truth, but it can only be rarely, and most frequently by accident.

CHAPTER IX.

Dislocations of rocks, or faults, constituting lines of valleys are by no means confined to mountain-chains, though more strikingly obvious in them than elsewhere: they are also sufficiently abundant in those districts commonly termed hilly, in which a line representing a proportional section of the surface is undulating, not irregularly angular as across a mountain-chain. The amount of dislocation of necessity varies materially, whether produced by distance from the principal point of disturbance, or by differences in the intensity of the force which has caused the faults. The dislocations are sometimes on so large a scale, that it is impossible to overlook them, while at others they are either so small or concealed by vegetation and detritus, that it requires much care and observation to ascertain their existence. The most favourable conditions for observation are where the rocks are either horizontal or inclined at small angles; the country well traversed by rivers, or cut by a coast line; and the rocks of such marked characters, that any change in the position that they would occupy, if unbroken, is readily ascertained.

In many hilly districts, a more modern rock overlaps others of greater antiquity than itself, and the valleys cutting through the upper into the lower strata, thus exposing the latter, seem merely, to the unpractised observer, the results of a scooping action of water, capable of removing solid matter to a sufficient depth. The whole has often the appearance of an undisturbed mass of a more modern deposit resting horizontally, or nearly so, on inferior strata, the latter being brought to light by denuding forces, acting in particular lines. The Black Down Hills, on the confines of Devon and Somerset, are highly illustrative of such deceptive appearances. When rapidly viewed, they seem to be composed of horizontal, or nearly horizontal, beds of green-sand, which cover either inferior oolite, lias, or red marl (for the two former fine off beneath the general green-sand area of these hills), here and there supporting a few patches of chalk. The appearance of the inferior rocks seems merely due to the relative depths of the various valleys scooped out by the action of water. The section on the following page will, however, show that the rocks composing these hills have been fractured subsequently to their deposition, and that, as far as it extends, the valleys are lines of fault. These, though irregular in the small scale, have a general northerly direction when viewed relatively to the district in which they occur.

The annexed section (fig. 34.) is made along a line extending from about S.W. to N.E., nearly at right angles to the greater part of the valleys comprised within it, and which happen to have such directions along this particular line. Of all those here exhibited, the Widworthy Valley shows the least dislocation, and it is only by a very careful estimate of the relative heights of the green-sand on either side, with allowances for its inclination on the opposite hills, that we can feel assured that there is any fault. The rest require no explanation, except that at the fault

Fig. 34.

SECTION ACROSS A SOUTHERN PART OF THE BLACK DOWN HILLS.

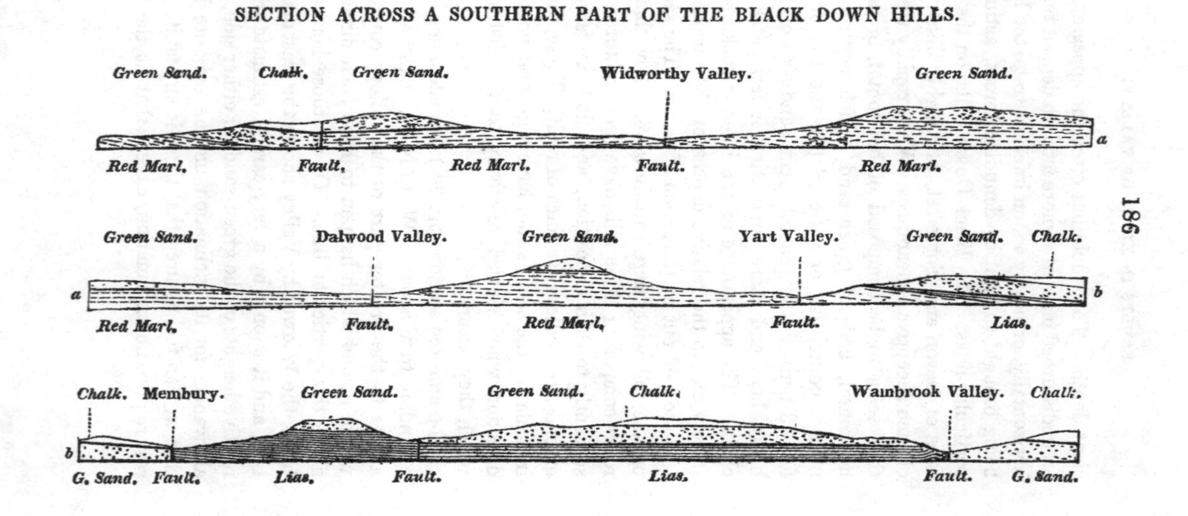

in the Wambrook Valley, where the lias abuts against the green-sand, the former is bent and contorted near the line of dislocation, in a far more remarkable manner than can be shown in the scale here adopted. This section is proportional, that is, the perpendicular heights and horizontal distances are upon the same scale; and hence the reader may form an idea of the real outline of that which would be generally considered as a remarkably hilly country, one where the traveller would describe abrupt slopes, deep valleys and high hills, the summits of those here represented, varying from 500 to 600 feet above the level of the sea.

That all faults are not necessarily lines of valley müst be well known to every practical geologist; indeed, in the section here given to illustrate the remarkable coincidence of the lines of several contiguous valleys with those of faults, it will be seen that there is no valley at the fault near Widworthy, where chalk abuts against green-sand. If the sectional line had been made to pass somewhat more to the southward, the green-sand would appear suddenly to rise in a boss above the chalk. The Wambrook fault, also, in its course to the northward, runs across the summit of the high hill named Coombe Beacon, dividing it geologically into two masses, the western side being composed of green-sand, and the eastern of chalk, while the outline of the hill continues rounded and unbroken, as if no dislocating force had ever acted across the rocks of which it is formed. Coal-miners, to whom faults are familiar, know full well that valleys do not always run in lines corresponding to them; even those in which the heave is ascertained by the workings to be considerable, often give no indications of their presence by depressions on the surface.

In thus showing that cracks or dislocations of strata frequently coincide with lines of valley, we merely intend to call attention to the numerous fractures upon which such lines, in undulating and hilly countries, keep those of faults, without by any means inferring that valleys may not be frequently due to the erosive action of water on strata which have not been dislocated, at least in those particular lines. Let the movement of water which has cut out the solid matter of the earth's surface into valleys, and transported the resulting detritus to places where its deposit has, geologically speaking, produced new rocks, be the effect of the greater or less intensity of any particular force, or due to any particular cause, preexisting cracks and fissures, more especially if they were fresh, would offer the lines of least resistance to an aqueous abrading power. Whether, therefore, valleys be due to the long-continued action of running water, or to the cutting power of water moving with rapidity in greater masses, we should always expect to find, as we do find, a very great coincidence between the lines of fault and those of valleys.

When we attentively regard really proportional sections of hilly countries, those in which the lines of surface are represented with a fair approach to accuracy, we find the depressions and elevations are generally so trifling, compared with the horizontal distances, that when we construct a model faithfully and proportionally representing the surface of a country which would be generally termed exceedingly hilly, such, for instance, as Devonshire, we can scarcely believe that we have not undervalued the relative heights, so small do they appear. Hence we are apt to consider such a model inaccurate, though constructed with every care and attention even by ourselves; so little does it seem to accord with our ideas of the elevations and depressions in the country itself. No doubt, if we attentively considered the subject, and properly estimated the relative heights and distances, we should be ultimately convinced that the model in question was a real miniature representation of nature; but all our previous ideas respecting the particular district having been formed with reference to our own height as a standard, and generally viewing the slopes of hills in such directions that they appear far more steep than they really are, there is greater difficulty in properly estimating the proportional undulations of a hilly country, than those who have never seen a true model of one would easily credit.

It has sometimes been supposed that existing marine currents or streams of tide are sufficient to cut out similar valleys. When, however, we turn our attention to what is now going on in situations most favourable to the cutting powers of tides or currents, namely, around coasts, where their velocities would be greatest, we do not appear to discover any facts to warrant such a conclusion; neither does it appear possible for the force of any existing marine current to produce the effects observed. The soundings round coasts present us with no lines which we might consider to be those of valleys, but with extensive areas, which would, if raised above the level of the sea, form great plains, with here and there some minor elevations and deeper depressions, the latter generally in the form of basins. There are, indeed, some long trough-like cavities in the North Seas, named the Silver Pits, but they present us with nothing like a system of valleys resembling those on dry land*. If the British Islands were elevated one hundred fathoms above the level of the ocean, and thus joined to the continent of Europe, they would be surrounded by an extensive area of flat land; for the fall from the coasts to the new sea-coast would be generally so gradual as to present to the eye one great plain.

The map opposite will show the relative amount of area which would become dry land by this comparatively small amount of change in the levels of sea and land. The unshaded parts constitute the present dry land, the lightly shaded portions the area which

* I am indebted to Captain Hewett, R.N., whose beautiful charts of the North Sea, not yet published, so well exhibit the forms of these singular lines of pits or troughs, for information which leads me to suspect that the Silver Pits are cracks or fissures resembling those termed valleys of elevation. As a general fact, the bottom rises on either side to the edges of the pit, where it suddenly descends from the depth of a few fathoms to 40 or 50 fathoms, the interior escarpments being very steep. In one place there is a rise of land in the central part of the trough, so that a section of the Silver Pits in their deepest parts would give an outline exceedingly like those of valleys of elevation, so well known to geologists, and produced by an upward break or crack often on the top of a great bend of strata. There is a curious line of bank between the Silver Pits and other lines of bank, parallel to themselves and the first line, on the S.E. of them, the whole having much the appearance of being the result of some disrupting and elevating force acting in a line somewhat N.W. and S.E.

The reader will not fail to observe, in the opposite plate, the general S.W. and N.E. line on the western edge of the hundred-fathom line, coinciding with the general range of the older strata in many parts of our islands.

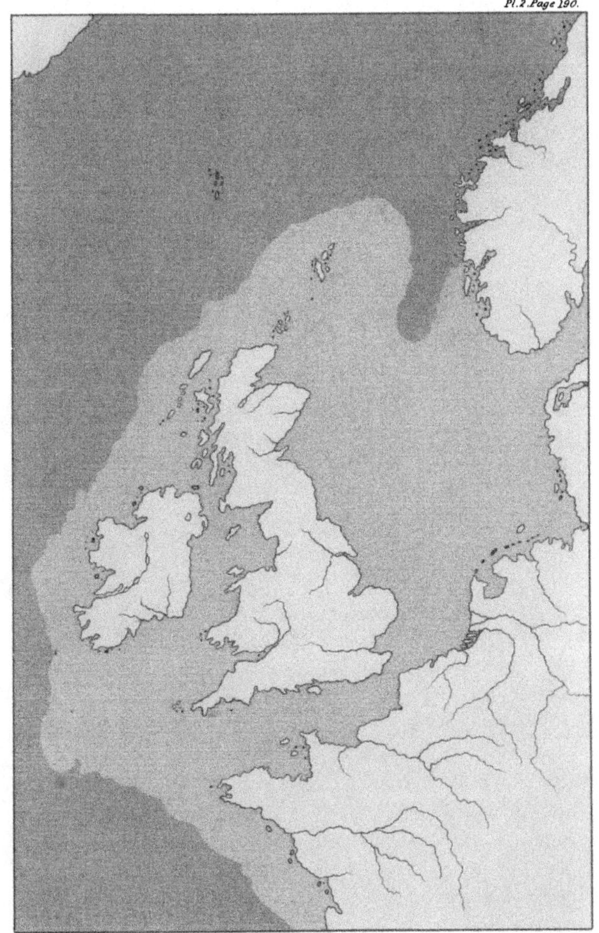

Engraved by J. Gardner

would be raised from the bottom of the present sea into the atmosphere, and the darker parts the portion that would still be covered by the waters of the ocean. The reader may judge of the extent of this land, thus brought above the level of the sea, by comparing it with the annexed square (fig. 35.), which represents

Fig. 35.

□

an area of 1000 square miles upon the same scale. Throughout the whole of the former soundings, thus raised above the sea, there would be no system of hills and valleys corresponding with those on the previously dry land. There would be elevations and depressions corresponding with various sand-banks, and some pits or hollows; but when we construct proportional sections of the general surface, even on a large scale, we find that these inequalities are generally trifling, with the exception of the Silver Pits above mentioned, though, from the necessary notice taken of them in charts, we may attribute an importance to them which they do not possess. It may here be observed, that if, instead of taking the line of one hundred fathoms, we had selected that of two hundred fathoms, the amount of area in that case added would be relatively inconsiderable, as the latter line is not far outside the former. Thus, after keeping a surface which would in general terms be called a great tract of plain, the area under consideration, bounded on the W. and N.W. by a line including the British Islands, and passing round them from the coast of France to that of Norway, would terminate seaward by a slope, sinking in a comparatively rapid manner into deep water.

Now all this extensive area, within the line of one hundred fathoms, is placed under conditions highly favourable to the production of valleys by the erosive action of streams of tides and currents, if the latter could produce the effects required; for the depth being inconsiderable, the bottom is more readily brought within their influence, without taking into account the increased velocities of each produced by coasts or banks. The action of the waves would certainly tend to equalize disturbed matter, particularly at moderate depths, and would thus constitute an antagonist force to the erosive action of streams of tides or currents: but this only shows that the conditions are unfavourable to the production of valleys by such means. It will, we conceive, be readily admitted that the deeper the water the less the power of streams of tide or currents to act on the bottom; therefore we may fairly conclude that neither in deep nor shallow water could valleys be produced, resembling those observable in hilly countries, by streams of tides or currents possessing no greater velocity than those known to us.

When we study the soundings along a coast line, we find that the valleys on dry land are rarely continued under the seas, unless in situations where mountains plunge suddenly into the water. In general, lines of valley terminate abruptly when they meet the sea, as should happen when coasts are cut back; for the tendency of the waves and streams of tide in such situations is to level inequalities, and produce a slightly inclined plane surface. The same effects would be produced on coasts not cut back to any great extent; and also where valleys are continued from the land under the sea, though they would not be apparent until after a long lapse of time, one depending on the depth of water on the one hand, and the amount of detritus carried round the coast and distributed by the action of tides, currents and waves on the other.

The island of Corsica may be taken in illustration both of sub-marine valleys and of flat bottoms near coasts, observable in a comparatively moderate area, and is the more striking because the great distributing power of tides in such situations is wanting. On the west coast the land plunges suddenly into the sea, and the valleys are continued beneath it, presenting numerous inlets where the level of the sea meets the inclined bottom of the valley. Soundings appear to show that the detritus borne down these valleys, or derived from the coast itself by the action of the waves, is accumulated beneath, and that there is a tendency to fill up the sub-marine valleys by this transported matter. If Corsica had been situated in a sea where tides and currents swept round the island, the filling up of the sub-marine valleys would have been more perfect, and the resulting deposits more horizontal. On the east coast of the same island the slopes are moderate, and the bottom of the sea on that side presents a more even surface, while the coast line itself is not broken into those irregularities so common on the western side. Detrital matter is moreover now forced on this shore by the action of the waves, forming lines of beach, which produce lakes behind them, such as those at Biguglia, Urbino, &c. On this side we do not discover sub-marine valleys. It will readily be seen that the sub-marine valleys are merely the continuation of those on the adjoining dry land; they, consequently, have not been produced by the erosive action of the sea upon them: so far is this from being the case, that where the action of the sea can produce any appreciable effect, inequalities are levelled. It will be obvious that the waters rushing down the inclined valleys of the western coast would lose their velocities, and, consequently, their transporting powers, when they entered the sea; so that far from producing a scooping effect on the sub-marine valleys, their action would tend to fill them up, by carrying detritus into them.

The sub-marine valleys observable off the steep coasts of a tideless sea, such as the Mediterranean may be termed when such subjects as these are considered, could not have been produced in the sea itself; they must have been formed by the same causes to which their continued higher or sub-aërial portions are also due. If the whole has been raised from beneath the sea, we do not perceive how the valleys were originally produced, at least by any scooping powers similar to those now existing in the Mediterranean; they may, however, have been formed in the atmosphere, and have been subsequently depressed beneath the sea. So far from sub-marine valleys along lines of coast exposed to tides and currents being more favourably situated for erosion than those observable in tideless, or nearly tideless, seas, they would have a tendency, all other things being equal, to be sooner filled up, and a plane surface the result, the transporting power of the tides or currents being the principal agent by which detritus would be brought to the required situations. With regard to all the sub-marine valleys yet noticed, and observable off mountainous or steep coasts, such valleys are mere continuations of those on the land, and hence must have been formed by the same general causes, and those causes not the tides and currents now existing in the same localities. These do not erode the surface beneath them in a similar manner, but, on the contrary, produce tracts of nearly level matter, such as occur extensively off some coasts, and which, if elevated above the level of the sea, would form broad tracts of land, with a surface sufficiently flat to be termed plains.

If actual streams of tide and marine currents possessed sufficient cutting power, and, consequently, sufficient velocity, to produce valleys, such as we observe in most hilly countries, we do not see how those fish which are found round most coasts could exist: the water would be charged with detritus, amid which they would be hurried away, or forced to use extraordinary exertions to retain their places. They might indeed avail themselves of eddies; but they would there be mixed up with the transported detritus, tending to come to rest in such places, so that their actual habits and those under such conditions would be widely different. If actual streams of tide and marine currents cut out valleys, we can scarcely see how molluscous creatures could survive; for the velocity required to remove hard solid rocks would not permit mud, sand and gravel, the retreats of many of these creatures, to remain; they could only exist in a few favourable situations. The formation of valleys, similar to such as we observe in hilly countries, by the erosive action of actual streams of tides or marine currents, is so opposed to observed facts, and would be productive of such consequences, that we are somewhat at a loss to conceive how such an hypothesis could have originated.

As streams of tide and marine currents, such as we now observe, appear incapable of producing those systems of valleys so commonly observable on dry land, we seem compelled, in our endeavours to explain the phænomena observed, to have recourse either to the long-continued and erosive action of torrents, rivers and minor streams of water, or to masses of water thrown into violent action and moving with considerable velocity. When we attentively consider the systems of valleys in a hilly country, one in which several varieties of rock occur, as very frequently happens, we find, viewing the subject generally, that hard rocks are rounded as well as those we should consider more easily disintegrated. We are therefore led to consider that there may have been some decomposition of the rocks acted on, which should, to a certain extent, render them all more equally removable by moving water. As far as our knowledge extends, rocks situated beneath the sea are not subjected to such general decomposing conditions as those exposed to the variable action of the atmosphere. The atmospheric air contained in waters, to at least certain depths, would no doubt furnish oxygen for the production of oxides, some of which might render a rock more easily acted on by moving water; yet decomposition, produced by atmospheric influences, would, we should suppose, be far more considerable in a given time.

There are few rocks exposed to what have been termed meteoric influences, or the general action of atmospheric causes, which have not suffered decomposition to greater or less depths*. In some it is very remarkable, so that there is a difficulty in procuring portions of such rocks for economical purposes, except at considerable depths, those on the surface easily splitting into fragments, even though apparently solid when first removed from the quarry. When deep cuts are made beneath the surface of hills, for new lines of road or other purposes, excellent opportunities are afforded for studying this decomposition, which is frequently seen to be irregular, as in the section beneath (fig. 36.), one not uncommon

Fig. 36.

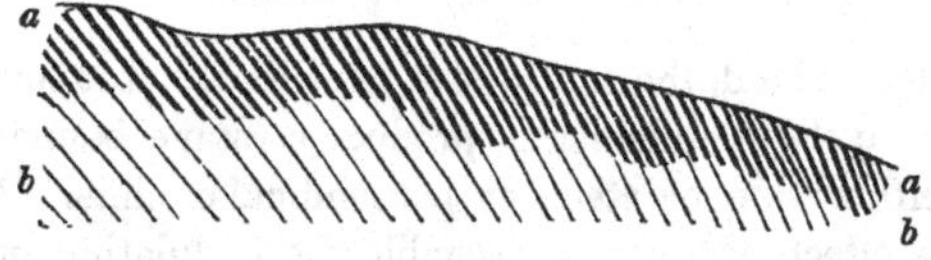

in some of the deep cuts for new roads among the argillaceous slates of Devon. The darker shaded portion *a a* forms the continuation of the laminæ of slates, represented in the lighter portion *b b*, the uneven intermediate line being generally well shown from the iron in the upper portion having been oxidized to a greater extent than that in the lower part, so that the upper portion is brown, or reddish brown, while the under retains its original grey tint.

When we attentively regard the manner in which rivers or torrents act upon the surface of a hilly

* Geological Manual, p. 43.

country, we observe that while much matter is removed by them, the resulting surface is not rounded, though minor little streamlets, often very destructive, may cut more or less deeply, and thus by their numbers and united action produce a smoother general surface than would be afforded by the erosive action of torrents or rivers alone. It will be seen that these lines of running water, by whatever names they may be designated, have a tendency to cut perpendicularly downwards, leaving steep or perpendicular walls on either side. If *b b* (fig. 37.) represent the continuous

Fig. 37.

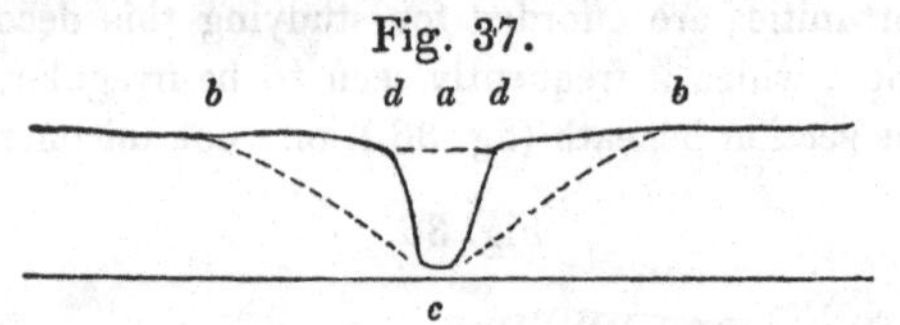

surface of land, the action of a river would, if it could cut into the rock beneath, produce a ravine bounded on either side by steep or perpendicular cliffs. As these effects would not resemble the undulating surface of a hilly country, and as the river would tend, if levels were favourable, still to cut down perpendicularly, we have next to inquire how far the decomposing action of the atmosphere may assist in producing the forms required. There would no doubt be a tendency in the summits of the cliffs, *d d*, to crumble and fall into the ravine (*a*) beneath, where the river would as constantly endeavour to remove the fragments thus thrown into it. Whether it could, or could not, remove all the fragments would depend upon its velocity; but in either case we might obtain slopes on both sides, if time sufficient were allowed.

We will here suppose the most favourable state of things, and that the river could carry away the fallen fragments, and that in consequence it ceased to cut perpendicularly to any perceptible extent, its force being employed in removing the new obstacles opposed to its passage. We might thus eventually obtain the slopes *c b*, *c b*, (fig. 37.) from the united action of gravity and atmospheric influences. Assuming that the slope thus obtained may be so smooth that, when ultimately covered with vegetation, no very striking unevenness should be apparent, we have next to discover the extent to which this sloping process may be carried so as to resemble the sides of hills in such districts as those under consideration. When the angle of inclination became $= 45°$, there would be a tendency of the fragments to retain their places from the support afforded to each other; so that further degradation would be principally due to decomposition, and removal by running water, the friction of the latter, particularly when charged with detritus, also producing its destructive effects. We should now obtain furrows and other inequalities from the cutting action of the streamlets, and there would be no smooth surface, if the slope became covered by vegetation, as we should from all analogy expect it to be. If it were not so covered, there might be much crossing of small streamlets or rills of water, and a moderately uniform surface the result.

Having arrived at this point, there is considerable difficulty in seeing how the slopes of the hills should become so gradual, from a repetition of such causes, as to produce the present appearances in a hilly country. There is generally a small depth of detrital matter above the fundamental rock on either side of the valleys, with a larger amount of the same matter at the bottom; but this superficial detritus is so far from receiving additions by the present action of atmospheric causes, which ought, according to this hypothesis to be only a repetition of that to which the valleys themselves are due, that gulleys with steep or perpendicular sides are cut through this upper detritus, and the rivers in the bottoms of the valleys also cut into it. We thus appear to arrive at the conclusion, that though valleys may be formed after a great lapse of time by the action of running waters on lines of fissures or depressions, we do not eventually obtain such comparatively gentle slopes as are common in hilly countries.

We should not here neglect the fact, often insisted on, that the windings of rivers in valleys, and their frequent changes of course, would cause levelling effects. Undoubtedly the flatter the surface the more irregular the windings of a river, all other things being equal. Hence it may be said that the serpentine windings would become less, the deeper the river cut into the valley; so that there would be no necessity for supposing the rivers, when they first acted on lines of valley, to have been far greater than at present, and that they have since dwindled down to those we now see. The latter is an hypothesis which would imply not only vast bodies of water rushing down with sufficient force to excavate valleys, but such an abundance of rain to supply the rivers, that the amount of that which now falls in the tropics would be a mere trifle when compared with it. The winding of rivers may have had an influence, and though probably not to the extent supposed, yet productive of effects which might afford assistance in explaining certain appearances in broad shallow valleys.

We have hitherto considered the rivers to possess a sufficient body of water, and to move with considerable velocity, so that they may be able to cut deep into the land, and produce the cliffs which by their disintegration should afford slopes. It is only necessary to traverse any moderate amount of hilly country to see that the waters now flowing in them, even with every allowance for freshes or floods, would be unable to produce the effects required. In many valleys the drainage waters are mere rills, and can never rise into importance, even in floods. In others there are no running waters, as may readily be seen in various parts of our own country, the water derived from the atmosphere being readily absorbed by the rocks beneath, either from their being porous, or from the edges of their laminæ or strata, as the case may be, sufficiently approaching a perpendicular position.

When from a lofty isolated point we cast our eyes over a considerable extent of hilly country, or view a correct representation of that country either in a model or a map, we are led to consider the inequalities before us as the effects of several causes, not the result of one alone. The drainage waters find their way by numerous channels to the principal rivers, uniting into various little systems of minor streams before they accomplish this end. Here and there little flats show that water has been arrested by barriers beyond them, and that the lakes thus produced have checked the velocities of the descending waters, so that detritus has been deposited, and the basin filled up*.

When we now reflect that, in all probability, the land we contemplate is traversed by fissures in various directions, some occasionally predominating more than others, we are led to inquire if all this dislocation of strata would not produce motion in the waters of a sea, if the rocks so fractured were broken up beneath it. When modern earthquakes are violent, great waves are driven on the land, where they often sweep all before them. We may therefore fairly infer that greater intensities of force would produce greater waves, while the motion caused in the water over the dislocations, supposing such to be formed, would tend to remove the fragments produced. The waves no doubt would act powerfully where they broke, and tend to scoop out the broken portions of faults or fissures; but we can scarcely obtain in this manner, or by the action of violently agitated water above lines of dislocations, such valleys as those observable in a hilly country. When we call in the aid of large masses of water moving with great velocity over rocks fractured and upturned in various manners, we certainly seem to attain a force capable of producing that general rounded character so remarkable in hilly land. The more sudden the rush of water, and the greater the volume, the more powerful would be the grinding and rounding action on the projecting points.

* In such a country the drainage is generally perfect, lakes being commonly found among, or on the skirts of, mountain ranges, or in extensive planes, so that great length of time is at least one of the causes which has produced the effects we should, in this case, see before us.

If we suppose such a moving mass of water to traverse land furrowed into preexisting valleys, the surface of which had been previously long exposed to the disintegrating powers of the atmosphere, we seem likely to obtain effects more corresponding with the facts observed, than if it rushed over land not so circumstanced. It has been above observed that hills and valleys, affording an undulating outline, are found as well among hard as among softer rocks. Now there appears a difficulty in obtaining similar valleys in rocks differing so much in their resisting powers to moving water, either by rivers acting on fissures, by sudden rushes of water or dislocations, or by their combined action, though much might be accomplished in the latter case. When, however, we add the decomposition and disintegration of rocks from meteoric influences, we seem in a great degree to remove the difficulties; for many rocks, such as certain granites, trappean rocks and limestones, which, when not disintegrated or decomposed, powerfully resist the action of moving water, would readily give way if reduced to fragments or soft yielding substances, as frequently happens by the action of the atmosphere upon them. Indeed they would frequently be the rocks most easily removed by the sudden rush of a great body of water.

The strata of a hilly country are often as broken, contorted, and heaved out of their places, though not to the same amount, as in mountain-chains. The angular outline of a system of mountains is, however, so different from the undulating wavy line of a hilly country, that we are led to suppose some cause to have acted on the one which has not acted on the other, or having done so, has not produced the same rounding effects. It has been sometimes considered that when we suppose water to have rushed over land in large masses, we call in the aid of some force almost preternatural, some tremendous catastrophe difficult to conceive. If, however, we look at the causes which may produce great debacles or deluges a little more on the large scale, we shall find that such effects are not very difficult to conceive.

Earthquakes, which, from the extent of the areas shaken by them at the same time, seem a modification or less intense action of the same force that has broken up rocks into fragments of various sizes, cause, by the vibration communicated to the waters of the ocean, waves of various heights to rush upon the land. The vibration of the sea produced by the great Lisbon earthquake of 1755 threw a wave sixty feet high on the coast of Cadiz, and one eighteen feet in height on the island of Madeira. The area agitated by this earthquake comprises a large portion of the northern hemisphere, yet no very great amount of dislocating effects was observed, at least nothing like the production of a line of elevated land. We may hence form some idea of the effects which would be produced if a line of mountains one hundred miles long, and not above two or three thousand feet high, were suddenly thrust up beneath the waters of the sea. The vibrations produced in the superincumbent fluid would be proportionally great, and the waves rushing over shallows and low lands comparatively enormous. If those who consider such an uprise of mountains a great exertion of force difficult of conception, were to take a common globe, about a foot in diameter, in their hands, and then estimate the proportional length of the crack, and the relative elevation necessary to produce not only this but far greater effects of a similar kind, they would probably cease to regard them as at all wonderful. Whether we suppose mountain ranges to be produced by the squeezing of the sides of fissures against each other, or by the intrusion of igneous matter, such an elevation as that noticed above would be inconsiderable when viewed with reference either to the mass or superficies of the earth.

CHAPTER X.

Faults, as dislocations of strata are commonly termed, being mere fissures, the sides of which have been more or less moved, are necessarily exceedingly variable with regard to their appearances. The walls or sides of a fault are sometimes so close, that there is a difficulty in introducing a small wedge into the fissures, while at others they are broad and filled with rubbish, generally fragments of the rock on either side; and when these are formed of various beds, such faults may contain broken portions of them. We cannot conceive a fracture of rocks perfectly smooth, on the contrary we should expect it to be uneven; and therefore, when the sides of a fault are found to be tolerably smooth, we should infer that the rubbing or grating of the sides against each had caused such appearances. Now when we attentively examine the sides of such faults, we very generally discover a polish and striated marks, indicative of pressure and motion, precisely as we should expect if they had been compelled to rub against each other with great force. These appearances are often seen in metalliferous veins, and are then known by the name of *slickensides*. There appears, however, to be little essential difference between slickensides, which are either smooth metallic coatings or the polished surfaces of a mass of metal, and the polished sides of faults. Striated parts are common in both, and it not unfrequently happens that in some metalliferous veins and common faults there are fragments of rock jammed in between the sides of the vein or fault, which appear to have been squeezed and polished in those positions.

It must necessarily happen, from the unequal fracture of rocks producing a fault, that the motion of the sides will bring some portions into contact while others will have spaces between them. Let the annexed diagram (fig. 38.) represent a horizontal section of a fault:

Fig. 38.

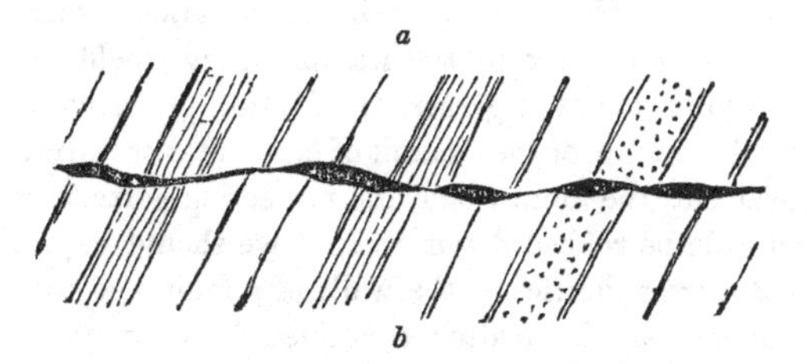

the side *a* being slightly moved to the left, *b* remaining firm, we have portions touching while others are distant. Now such appearances are common, and the analogy between them and metalliferous veins highly interesting. The thinning and thickening of such veins may be easily illustrated, if the reader will take a piece of writing-paper, divide it into two by a waving line, then place it upon a piece of gilt paper, and slide one portion of the divided paper so that the waving lines of the edges do not coincide, and he will perceive the gilt paper between the interstices, resembling a metalliferous vein, running with unequal thickness, through the upper paper. This process has been adopted in the drawing for the above wood-cut, white paper having been divided and moved on a black surface*. It will be obvious that a very moderate motion, either lateral or perpendicular, or one compounded of both, would cause such appearances as these, which are commonly observable in metalliferous veins, and in those faults where the labours of man render them better understood than they can be by natural sections at comparatively distant places.

In thus endeavouring to explain the uneven character of the sides of faults, we must not suppose that if again brought to those relative situations where they occurred prior to the fracture they would exactly coincide. The greater the friction produced by lateral pressure, or the amount of horizontal or vertical movement, the more would the projecting portions of each side be rubbed down, so that we should expect greater irregularities in the walls of a fault, and consequently in the matter contained between them, when the dislocating movement had been small or moderate.

Lines of fault are commonly described as straight, or nearly so, and certainly some do keep remarkably straight courses for considerable distances if measured by yards or fathoms; but they more frequently run in a waved or irregularly broken manner, precisely as we should expect would be the case with fractures.

* A still more illustrative mode of exhibiting the form of matter required to fill up the interstices of a fault with irregular sides, is by procuring an uneven fracture of a mass of plaster of Paris, sliding one of the surfaces upon the other so that numerous interstices occur, and then cast lead into the cavity, having walled up the sides to prevent the escape of the liquid metal. By cutting the resulting piece of lead in various directions, we obtain sections strongly reminding us of the observed inequalities in metalliferous veins.

Though many keep a remarkably general course, often parallel or nearly so to each other, when viewed with reference to a large area, the smaller parts of such courses are often very irregular. We may illustrate this by the following diagram (fig. 39.), representing a fault, on the north of Weymouth, traced with much care for about thirteen miles, and laid down on the Ordnance map of the district, by Professor Buckland and the author*.

Fig. 39.

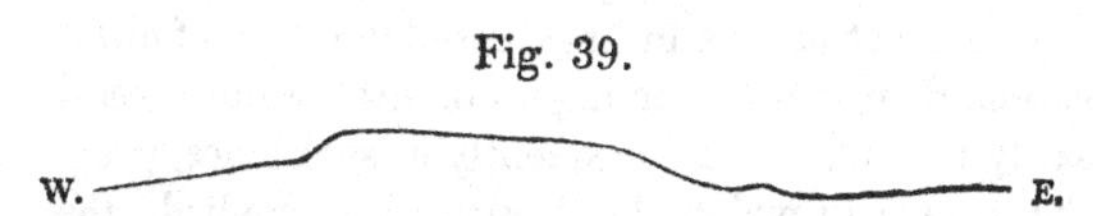

It will be observed that though the fault may be described as ranging, taken as a whole, in an east and west line, the minor parts vary, as might be expected, in their lines of direction. We should anticipate that similar facts would be observable in mountain-chains; with this only difference, that in lines of great fracture the length of the minor irregular portions would often be considerable, though still short when compared with the whole length of the chain. Hence instead of regarding the irregularities of strike or direction in the rocks of a mountain range as proofs that no general line of contemporaneous elevation prevails, we should take the general bearing of the whole, precisely as we do that of a fault, carefully estimating, without bias, or determination to find things as we suppose they must be from preconceived opinions, how much dislocation may have been

* See Geological Transactions, 2nd series, vol. iv.

produced at one epoch, and how much at others, should there be evidence, which there frequently is, of more great dislocations than one.

Faults sometimes divide into two or more lines, precisely as we should expect fractures to do; and when this happens, we frequently find the lines of dislocation approaching their termination. The sides of a fault occasionally exhibit flexures, showing that the rocks were to a certain extent in a yielding state, and that there has been pressure. As these contortions are not always in lines parallel to that of dislocation, though the doubling up of strata would necessarily take place most frequently in such lines, (from the manner in which the pressure was applied,) the various twists and flexures are sometimes cut by the general horizontal level in such a manner as to exhibit curved lines, illustrative of those greater curvatures and circular arrangements of strata described as circuses or circular valleys of elevation. These circuses or amphitheatres, in which a range of elevated land, composed of some one kind of rock, passes in a circular manner round a depression, with sometimes a central boss or elevation of land, may occasionally be seen well illustrated at the termination of small faults in nearly horizontal rocks, composed of unequally indurated strata, such as the lias, and laid bare by tides or other causes.

Fig. 40.

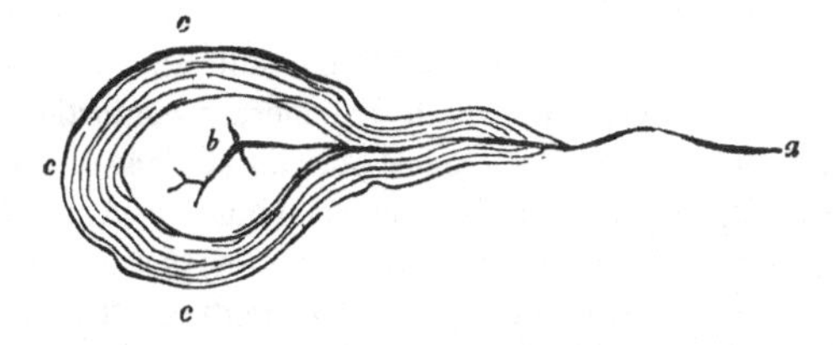

The above diagram will convey an idea of one of these small faults, *a*, which terminates at a boss, *b*, resulting from the flexure round the end of the crack or dislocation. This curvature being denuded horizontally affords the circular line, *c c c*, upon the same principle that a piece of mammellated malachite, stalactite, or similarly arranged substance, would, when cut in particular planes, present a series of concentric circles. If now a softer rock exist between the outer part, *c c c*, and the central boss *b*, such as a marl or shale bed between two strata of lias limestone, and moving water act upon the whole, we obtain a central elevation, a kind of circular valley, surrounded on the outer side by an escarpment of hard rock. It will readily strike those who have observed such circuses, or circular valleys of elevation, on the large scale in nature, that the miniature representation of them observable at the termination of a small fault is in principle the same in all. The outlets are generally also through a ravine or gorge, the part where the contortion or flexure becomes a crack. A section across such circular arrangements of beds, whether on the great or small scale, would only differ in magnitude, the general features being similar*. When we consider how completely accidental the form of any particular contortion may be, we are not surprised at the various figures presented by valleys or amphitheatres of elevation, which are merely contortions or fractures cut by the general horizontal plane of dry land, and exposed to the various denuding changes to which that particular part of the earth has been subjected, since the necessary force produced the original fracture or contortion.

It will readily strike the reader, that these amphitheatres or circular valleys of elevation merely require the ejection of volcanic matter through them to become the much-disputed craters of elevation. If, upon the principle exhibited in fig. 40., volcanic rocks, such as large horizontal accumulations of basalt, trachyte, volcanic conglomerates, or other substances of the like kind, were forced up in a circular manner, we should have a crater of elevation. It merely requires a volcano to play through the central part of the circle to produce all required by Von Buch's theory of the craters of elevation. As it appears very difficult to thrust a horizontal mass of rocks upwards, particularly if there be a point more powerfully acted on than others, without producing corresponding points of greater elevation than the rest, or fracture of such points into crateriform cavities, it seems somewhat surprising that craters of

* The study of small dislocations in various rocks laid bare horizontally for some distance by low tides, more particularly if they be composed, like the lias, of beds unequally indurated, affords abundant information respecting the greater dislocations observable on the earth's surface.

elevation have been doubted. When we find siliceous sandstones curved and twisted in every possible direction, as we frequently see them, we infer that the particles composing them have had the power so to move among each other, that a comparatively soft and yielding state has been produced. Without this yielding condition, the contortions could not have been effected. We are, therefore, not surprised to find those simple and larger curvatures of sandstones, slates, limestones, and other stratified rocks, which, when denuded in a horizontal plane, constitute great curved or circular lines of strata. Why it should be considered extraordinary that great tabular masses of basalt, trachyte, or rocks of that character, should yield, or become soft, when their composition is more favourable for such a condition than that of the sandstones, does not appear very clear.

If an elevating force act beneath a tabular mass, whether of the common fossiliferous rocks, or basalt and trachyte, and either a part of the mass be less resisting than others, or the force act with greater intensity in one place than in another, a protrusion, or somewhat circular fracture, would, we should consider, be the result; one that can scarcely be considered extraordinary; but, on the contrary, one that it would be more extraordinary not to find. Among the contortions and fractures of mountain-chains, and disturbed countries generally, there is no want of strata so circumstanced that if volcanic matter pierced the central parts of the area, they would readily be termed craters of elevation. The most beautiful example yet noticed in the British Islands, of strata so arranged, is that lately discovered by Mr. Murchison, in the grauwacke series, at Woolhope in Herefordshire*.

It sometimes happens that the side of a fault turned upwards by pressure, is not that which we should expect to have been so circumstanced. The following

Fig. 41.

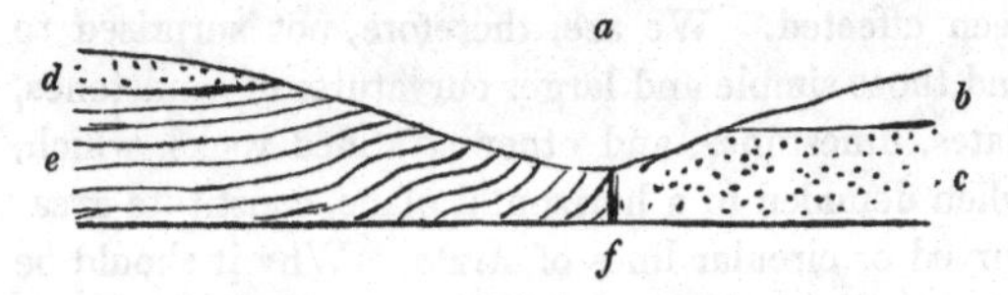

example is not selected because it is one readily seen in the country, for it requires some attention to be certain of the facts, but because it is a larger view of the Wambrook fault, figured p. 186, and there shown to be one of a system of faults. Among the faults on sea-coasts better examples may be found, because the facts are there seen at a glance, as if in a picture. By means of the fault, *f*, in the Wambrook valley, *a*, lias, *e*, on the west, is brought into contact with green sand, *c*, on the east, covered by chalk, *b*, and once constituting a continuous portion of the green sand, *d*, on the west. Now, if a simple dislocating force either produced the mere subsidence of the mass on the east, or the elevation of that on the west, the ends of the lias strata ought to be turned downwards instead of upwards as they now are, supposing the necessary friction. Either, therefore, there has been great lateral

* A figure of this valley or amphitheatre of elevation will probably appear in this gentleman's work on the upper grauwacke series of Wales and neighbouring parts of England, and on the geological structure of other adjoining districts.

pressure forcing the edges of the lias upwards, or, when the fracture was effected, the mass on the west was thrown upwards in such a manner as to descend against the eastern side of the dislocation or fault. In this case the bottom of the valley prevents us from observing whether the green sand bears marks of extraordinary pressure, as quarries are not opened on that side, but there is no appearance of any contortion down to the bottom of the valley. We cannot, therefore, infer here, as we may in many other places, where equally yielding rocks occur on both sides of faults, and one of them is turned up in the manner represented above, that a relative elevation of the one, greater than that now observable, once took place, and that either from gradual subsidence of that side since, or from sudden violence at the time of the fracture, the friction was so great as to turn up the ends of the relatively elevated rock.

The contents of common faults throw much light upon what have been termed metalliferous veins, merely because they contain metals sought for and employed for economical purposes. Many veins, which would not be commonly termed metalliferous, contain abundance of iron pyrites, often found in common faults, more particularly when they traverse clays, marls and slates. It has been above remarked that the sides of faults are frequently so close as absolutely to touch, while at times they gape, and the fissure has been variously filled. Angular fragments of different sizes, some no doubt broken off at the time of the dislocation, are common; but there have evidently also been hollow spaces, both among the fragments and between the walls of the faults themselves, which have only been filled gradually by the introduction of matter, principally in solution, and which in some cases has not absolutely filled the cavity, but left interior spaces, technically termed *druses*, surrounded by crystals of various kinds. The infiltration has often taken place precisely in the manner in which cavities have been filled in certain rocks, giving rise to the formation of agates.

It will generally be observed that the cavities or hollows in question are lined, if not entirely filled, with matter predominating in the rocks composing the sides of the fault. If these be of limestone, the matter is generally calcareous spar; if among siliceous grauwacke, or rocks of that character, quartz; if among argillo-calcareous beds, such as the lias, where sulphuret of iron may be common, calcareous spar, iron pyrites, and sometimes selenite, are not uncommon; if among the upper beds of the red marl, or other substances where sulphate of lime may be common, gypsum is frequently found*. There is evidently a connexion between the ordinary contents of the fault and the rocks among which it occurs, as we might expect, because it would readily receive water from them, and, consequently, any matter which it might hold in solution.

In some cases the fissure has evidently gaped for a long period, and the gradual filling up has proceeded from the sides inwards. The most obvious facts of

* The diagonal lines of flint traversing chalk, see *b*, fig. 15., p. 99., and having much the appearance of small cracks filled up by the infiltration of silica, seem also to afford an example of the separation of matter from the body of the rock, analogous to those above enumerated.

this kind are observed among limestones of a certain collective thickness, such, for instance, as the carboniferous limestones of England and Wales. The annexed diagram represents a portion of such a fissure,

Fig. 42.

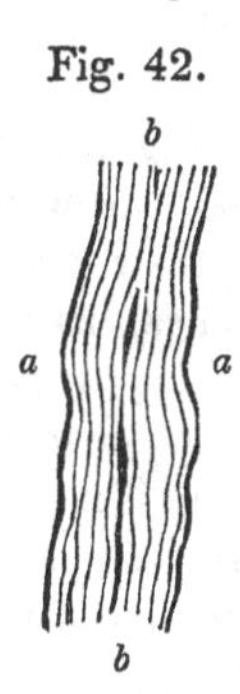

the curved and, to a certain extent, parallel lines *b b*, conforming to the sides *a a* of the fault. When great length of time has elapsed in the production of each coating of calcareous matter, the outside of each is faced with the heads of crystals, upon which the next coating has been formed. Cavities or druses are by no means uncommon in the central portions of such faults. Circumstances, precisely analogous, may be observed among various slate rocks, the matter coating the walls of the fault being quartzose instead of calcareous. Many of the faults, termed cross courses, in Devon and Cornwall are of this character.

The analogy between the common contents of a fault and the greater proportion of metalliferous veins is so striking, that we are irresistibly led to consider their origin to have been similar. It is only when we turn to the metalliferous contents that we have any apparent difficulty. The presence of the sulphurets of metals is not confined to the metalliferous veins, commonly so called; for sulphuret of iron may often be detected in common faults, clearly introduced into them subsequently to the fracture, and occasionally crystallized in the druses. The principal distinction seems to consist in finding particular metals in one place and not in another. Now, this being the case, we should rather study the conditions that may have produced such a difference, than conclude, *à priori*, that there can be no analogy between faults and numerous metalliferous veins or lodes. As well might it be contended that there is no analogy between two common faults, because the space between the walls of the one is filled with quartz, and that of the other with carbonate of lime.

Here some remarkable facts come to our aid. It has been very generally observed, and was more particularly insisted upon some years since by Mr. Carne, that the character of metalliferous veins changes with the structure of the rock through which they pass. There is scarcely a miner of any intelligence in a mining district who is not practically aware of this fact. He may often be heard to predict either success or failure from the change of country (as the rocks, including the veins, are technically termed in some districts,) discovered in following up a particular metalliferous vein. His predictions are not always verified, it is true, for the change may have been of a different character from what he anticipated; but this constant reference to probable change shows how common the fact is. It is not necessary that there should be an absolute difference of rock, geologically considered, such, for instance, as from slate to granite: a change in the hardness and general mineralogical structure of the same system of rocks will frequently cause alterations of the vein sufficiently remarkable.

Now this difference in the contents, or rather in part of the contents of metalliferous veins, clearly depends upon the rocks through which they pass; and if to it we add the facts above noticed, we have considerable evidence to show, that upon the rocks through which either metalliferous veins or common faults pass, depends the condition of the matter which may occur in them. Among the best conditions for expecting to find fissures filled not only with an abundance of metals, but also with kinds most useful to a civilized state of society, is the proximity of some mass of granite, porphyry, greenstone, or rocks of that class, to stratified rocks, such as slates and others*. Every Cornish miner knows the value of the proximity of his mine, if it be in killas, (as the slates are commonly termed in the mining districts of that part of England,) to granite. The best mining fields of Cornwall and Devon are within moderate distances on either side of the line separating the two rocks†. Here, therefore, we have a condition favourable to the occurrence of particular metals. Analogous facts

* M. Necker de Saussure has accumulated considerable evidence on this head in a paper read before the Geological Society of London. See Proceedings of that Society.

† It may be here noticed, as a curious fact, that the numerous situations in Devon and Eastern Cornwall, where bundles of manganese occur, the latter are, with very few exceptions, and many of those somewhat doubtful, in grauwacke, near greenstone or some trappean rock.

could be readily adduced, but notices of them would not accord with the plan of this volume. In many we should consider that long-continued heat had acted a considerable part in producing the phænomena observed, and not a few cracks or fissures might arise from the conditions under which an intruded mass of igneous matter, and the rocks through which it appeared, might eventually be placed.

That chemical changes have taken place in the matter contained in some metalliferous veins after they have, to a certain extent at least, been filled by various mineral substances, is proved by the occurrence of pseudo-crystals, or those of one substance formed in cavities left by crystals of other substances which had previously disappeared, their moulds or impressions remaining. The same circumstance is also well shown by the presence of the cavities themselves, left by the disappearance of the substances to which they are due. In the former case we have four different conditions: 1. the production of the original crystal; 2. the envelopment of them by a second substance, differing from the first, such as cubic crystals of fluate of lime by quartz; 3. the disappearance of the matter of the original crystals by solution or otherwise; and 4. the substitution of a third substance for the matter thus removed, by which it takes an external crystalline form it would not otherwise possess. In the second case there would be only the three first of these conditions.

These changes are highly interesting, as they show that the contents of a vein do not necessarily remain at rest after their first production, and prepare us for other important changes which may have been produced and are as yet unknown to us. The formation of one crystalline substance upon crystals of another substance also seems to show that the present contents of veins have not been suddenly produced, but have been the result of time. In metalliferous veins the ores of metals rarely extend from side to side: they are generally mixed up in various ways with quartz and other substances, and very frequently with fragments of rocks of the same kind as those through which the veins pass.

We have no intention of entering into the general subject of metalliferous veins, which would carry us into detail inconsistent with the plan of this volume; we are merely anxious to show how analogous many of them are to common faults, and that different conditions have produced differences in the substances filling the fissures. In some rocks metallic ores evidently form part of the mass in which they are contained, such as oxide of tin disseminated in some granites. It has merely crystallized out in the same manner as the felspar or mica. When also bunches of ore occur in the middle of a rock without fissures like faults, the formation appears somewhat contemporaneous. If we attentively consider many of those systems of strings or small veins of metal crossing each other in all directions, they appear like cracks produced in the rock during consolidation, and subsequently filled with metallic ores derived from the rock itself. The matter of the rock generally appears to have been once united; and many little marks may occasionally be traced on both sides of the string or small vein, which, if the latter were removed, would join. Other systems of a network of small veins appear as if they had separated out during the formation of the rock itself. But in this case, as in the other, the metallic matter filling the veins must have once been part of the matter of which the rock was composed. These veins appear exceedingly analogous to quartz and calcareous veins observable in many rocks.

In addition to the facts above noticed, which seem to point to the rocks in which metalliferous veins occur as the source whence much of the matter contained in them has been derived, it may be stated that the rocks themselves, where they approach the vein, are not generally in the same condition as at some short distance from it. We may very readily conceive that in metalliferous veins, more particularly those long formed, the contents may be variously derived. It would, however, be premature to reason on this subject before we have an abundance of data before us, and due attention be paid to the conditions under which any particular metalliferous vein or system of veins may have existed, carefully weighing those conditions, and observing how far the contents either differ from or resemble each other, according to the relative circumstances under which the veins may have been placed. By thus carefully separating and classifying the facts, we may be ultimately led to trace to their sources the causes, and the modification of the causes, which have produced them*.

* We have purposely abstained from noticing the probable effects of electricity in producing some of the contents of metalliferous veins, since the recent brilliant researches of Mr. Faraday have indirectly placed the subject in such a light, that it requires a careful examination of some hitherto neglected points before we attempt to reason on it generally.

CHAPTER XI.

After the remains of animals and vegetables, entombed at various depths and at different periods in the crust of the globe, were fully recognised as the exuviæ of organic life which had once existed on the surface of the earth, it became a somewhat prevalent opinion, particularly after the researches of Cuvier and Brongniart round Paris, and those of Smith in England, that contemporaneous deposits were characterized by the presence of similar organic remains. During the time that such deposits were supposed to be distinguished by similar mineralogical composition, and, viewing the subject the other way, that similar mineralogical structure at once proclaimed the geological date of the rock, it was considered somewhat heretical to doubt the possibility of discovering any other than a certain series of organic remains in a given fossiliferous rock wherever found.

This opinion, though somewhat modified, is still so far entertained that similar organic remains are supposed to characterize contemporaneous deposits to considerable distances; at least to this extent, that if a belemnite be discovered on the flanks of the Himalayan mountains, there is a disposition to consider, *à priori*, that it must belong to some part of a series of rocks in which this genus is found in Western Europe. Fossil shells of similar species are also supposed to characterize the same deposit over considerable areas.

In the present comparatively advanced state of geology, it behoves us carefully to weigh the conditions under which animal and vegetable life now exist, before we assume that a given deposit can or cannot be determined by its organic contents. Some of these conditions are sufficiently well known, so that by diligently studying the zoological and botanical characters of fossiliferous rocks, we may approximate towards ascertaining how far animal and vegetable life may have existed under circumstances resembling those we now observe, and how far both may have differed.

It is more particularly desirable to investigate the conditions necessary to marine animals, which, from the forms observed among organic remains, we consider to have been most abundantly entombed in mineral matter since the surface of the earth has been fitted for their existence; an abundance so great, that if their exuviæ were abstracted from the mass of fossiliferous rocks generally, a large amount of matter would be removed, and the relative volume of the whole be so reduced, that, in several instances, rocks composed of little else than the remains of marine creatures would almost entirely disappear.

The temperature of the sea is not liable to those great and sudden changes observable in the atmosphere. Variations in climate, according to the seasons, no doubt cause corresponding variations in the temperature of the waters beneath them; but the amount of change thus produced is relatively small, as may be seen by comparing the tables of the temperature of the sea on any given coast with those of the climate in the same situation.. Marine creatures therefore are not exposed to the necessity of change of place, from the mere effects of altered temperature, to the same extent as animals which exist in the atmosphere.

Now as the temperature of the sea is necessarily subjected to greater variations in shallow than in deep water, the creatures which habitually prefer the one to the other will (unless especially fitted to support variable temperatures, as no doubt a large proportion of them are,) be disposed to change their abodes more frequently in shallow water than in the deep sea, supposing it to be inhabited *. In other words, we should expect marine creatures, inhabiting shallow water around shores, to be less constantly found at given depths, than those which require a greater depth of water to render their abodes agreeable to them.

Taking the ocean-level at the equator as a line, and viewing the subject generally, the temperature would decrease perpendicularly both down to the lowest depths of the ocean, and to the highest parts of land which may project into the atmosphere. We should thus have certain heights or depths, as the case might be, where creatures preferring any given temperature might find it within certain limits. Temperature, however, though an important part of the

* Fish and molluscous animals are also disposed to change their places in shallow water, from the agitation caused by waves. Some seek refuge in deeper water, while others enter sheltered bays and creeks. Sometimes, if heavy gales prevail for a considerable time, marine animals will be found in estuaries which are entirely free from them in moderate weather. Pectens often enter into the creeks and estuaries of South Devon, if the water be not too fresh, when bad weather prevails in the open sea, to which they again return when the gales subside.

conditions necessary to the existence of animal life, and greatly affecting its distribution over the surface of the globe, is only one among many considerations to be taken into account in an inquiry of this kind. As we see that different densities of the atmosphere are preferred by different animals existing in it, so do we find various densities of water most agreeable to creatures existing in that medium. As we have elsewhere * remarked, if we consider that animal life decreases in proportion as the atmosphere becomes colder and less dense, and that marine life is less abundant as the pressure of the sea increases, and the necessary light diminishes, we obtain, as it were, two series of zones, one rising above the ocean-level, the other descending beneath it; the terms of the two series, all other things remaining the same, affording the greater amount of animal life as they respectively approach the ocean-level.

Every naturalist knows that a given depth of water is that at which certain creatures are likely to be found; which implies that they prefer a given pressure of water and particular temperatures. For instance, creatures fished up in the tropics from considerable depths, and never discovered in shallow water, exist habitually under greater pressure and at a less temperature than other tropical creatures found in shallow water. If we take any of the experiments made to ascertain the temperature of the sea at different depths in the tropics, we shall see how soon the creatures inhabiting the shallow water in such latitudes would be placed under circumstances by no means favourable

* Geological Manual, p. 29.

to their existence, neglecting for the present the variations of pressure which, no doubt, would produce equally destructive results. The experiments of Sabine, Wauchope and Lenz * show us that while there may be a temperature of about 80° Fahr. in the surface waters of tropical seas, there is a temperature of from 36° to 50° at the depth of about 1000 fathoms, the greater temperature being discovered nearest the equator. The manner in which this decrease takes place may be illustrated by the experiment of M. Lenz made in lat. 21° 14′ N. The surface water being at 79°·5, he obtained 61°·4 at 150 fathoms; 37°·7 at 440 fathoms; 37°·2 at 709 fathoms; and 36°·5 at 976 fathoms. It will readily be seen that the great change of temperature takes place beneath moderate depths of water, and that at considerable depths the relative amount of heat is more constant. This is a fact very generally observed, as might be anticipated, in other situations than the tropics.

The change of temperature by an increase in the depth of the ocean, resembling, to a certain extent, that caused by elevation in the atmosphere, naturally induces us to inquire whether creatures known to us as existing only in the waters of temperate climates might not be found at depths in the tropics where the temperature was nearly similar, upon the same principle that plants growing on the low lands of cold climates may be discovered on the elevated mountains of more temperate regions. It might be said that if pressure be taken into account, there is no reason why, *à priori*, certain plants should be more

* Geological Manual, p. 25.

capable of sustaining different degrees of pressure than certain marine creatures. We will not, however, stop to examine how far the cases would be analogous; for there is one very important consideration which must enter into all such inquiries as those we are now instituting, and that is, Would marine creatures living habitually in shallow waters, and consequently in a medium where atmospheric air was disseminated to a considerable extent, be capable of supporting life at great depths where we may infer that, if air be present, it would be infinitely more scarce?

We know, as far as regards freshwater fish, that if they be plunged into distilled water, they will die from the want of the air commonly contained in lake and river waters, and we may infer that marine fish could no more exist without air disseminated through sea, than the freshwater fish in distilled water.

The analogy between the plants and animals above noticed seems here to cease, for the plants could as readily procure the gaseous matter necessary for them in the one position as in the other; though it must not be forgotten that in the great adaptation of animal and vegetable life to the situations fitted for it, numerous plants which live in places where the density of the atmosphere is comparatively small, are, according to Humboldt, provided with an abundance of secreting vessels, so that the respiration of the leaves of such plants becomes deranged when the latter are transferred to situations where there is greater atmospheric pressure*.

* Tableaux de la Nature, ii. p. 115.

As we cannot readily conceive that marine creatures are capable of decomposing water for the purpose of procuring the oxygen that may be necessary for them, we look to the absorption of gaseous matter by water, and its consequent dissemination among the latter, as the means by which creatures obtain the requisite air, assuming that oxygen is necessary to the whole animal creation, and that it must be obtained at intervals, however unequal these may be in different creatures, for the purpose of sustaining life*. Of the depths to which air may thus find its way we know nothing; but the very curious observations of M. Biot on the gaseous contents of the swimming-bladders of fish show us that such contents probably vary, according to the depth at which the fish usually live. He found that such bladders were not filled with atmospheric air, but with nearly pure nitrogen, when the fish inhabited shallow water, and with oxygen and nitrogen, in the proportion of ·9 of the former to ·1 of the latter, when they inhabited depths of from 500 to 600 fathoms†. We might hence infer that there was a difficulty in procuring nitrogen at great depths while it was readily obtained near the surface, and hence that atmospheric air was more abundant in the

* It may, indeed, be said that if the lower creatures do consume oxygen forming part of the atmospheric air disseminated at great depths, the intervals at which they require it may be so long, and the amount so small, that a very trifling relative volume of this substance would suffice for great lapses of time. This latter view, if true, would, however, only tend still further to illustrate the adaptation of animals to the conditions under which they are placed.

† Biot, as quoted by Pouillet, Elémens de Physique Expérimentale, tom. i. p. 187.

latter than in the former case. We may also suppose that oxygen is more readily absorbed by sea-water than nitrogen, and that therefore it may extend to greater depths. Be this, however, as it may, the differences in the contents of the swimming-bladders is exceedingly remarkable, and apparently point to a difference in the gaseous matter disseminated through sea-water at various depths, at least as far as regards the proportions that oxygen and nitrogen may bear to each other under such conditions.

It might be considered that fish, being provided with these swimming-bladders, would have the power of rising to any height in the water they desired, and that therefore they could obtain whatever amount of disseminated air they might require. It would appear, however, that though capable of rising and descending from and to certain depths, fish are limited, according to their species, as to the thickness of the stratum of water, if we may so speak, which they inhabit. As fish rise and descend in water, at least the great proportion of them, by means of expanding and contracting the gas in their swimming-bladders, it is evident that when this gas becomes, by pressure, of the same density as the surrounding water, the fish can no longer descend, unless, indeed, by great muscular exertions; neither would they ascend with facility beyond a certain height. M. Pouillet has observed on this head that the gas in the swimming-bladders of fish brought up from the depth of about 3300 feet, and therefore under a pressure equal to about 100 atmospheres, increases so considerably in volume that all muscular effort being unable to restrain it, it forces the bladder, stomach, and other neighbouring parts outside the throat into the form of a balloon-shaped mass*.

We have no reason to conclude that different species of fish are the only marine creatures likely to be limited to particular depths of water; we should infer that all the animal inhabitants of the ocean and seas were similarly circumstanced. Pressure and temperature change with the depth; and we cannot conceive that the same animal, be it what it may, could live equally well near the surface and beneath a depth of 1000 fathoms, more than we should consider that a man would find himself equally comfortable at small heights in the atmosphere, and at an elevation of 30,000 feet above the sea. Throughout animal life creatures appear to have been formed to sustain that particular pressure, whether of air or water, which was usually found in the situations destined to be occupied by them. Creatures existing in the atmosphere, all other things being the same, would, however, suffer less in proportion from a given amount of vertical change of place than creatures living in the sea. An eagle, accustomed to soar at considerable heights, can live at the level of the sea; but it is very doubtful if a shark could continue long to exist beneath considerable depths, though no doubt rapacious creatures are better enabled to sustain such changes than others; the necessity of finding prey at various altitudes or depths, as the case may be, being especially provided for.

* Elémens de Physique Expérimentale, tom. i. p. 188. Seconde Edition.

Fish could not remain at ease in any particular depth of water if their specific gravity at the time was not precisely that of the medium in which they existed. It is evident that in many the change in their relative specific gravities is effected by the contraction and expansion of the gas in the swimming-bladders above noticed; and we have seen the consequences of removing such an elastic body from greater to less depths. We now come to another question. As the liquid matter circulating in marine creatures is, in all probability, of such densities that the pressure of the surrounding aqueous medium is exactly such as to allow it to move with the proper freedom, would change in the relative depth of any given marine creature produce a corresponding effect in the circulation of its fluids? Although we may consider the fluids and the surrounding water to be to a certain extent elastic, their elasticity is not such as to produce effects similar to those caused by the expansion of gaseous matter: it would probably, therefore, require very considerable differences in pressure to produce any appreciable effect.

Wherever a creature exists, we may consider it to be partially kept together by a given amount of pressure, so arranged that it can move freely in the medium, whether gaseous or aqueous, in which it may habitually live. The powers of the muscles are suited to this condition of things. If, however, the pressure become less from any cause, there would be a tendency in the body of the creature to change its volume in order to accommodate itself to the medium by which it is surrounded. The derangement would be first experienced in the smaller and more delicate vessels, constructed as they are with such beautiful precision, that the power of controling their action is always equal to the effects required. A pressure suited to the proper condition of the animal is part of this power; and when it is greater or less than the creature habitually supports, the animal suffers in proportion to the difference; so that when the latter becomes considerable, the creature ceases to exist. Although man can live without unpleasant sensations under a considerable variation of atmospheric pressure, yet upon lofty mountains, such as Mont Blanc, and, consequently, under a considerably diminished weight of air, he perceives he has attained a situation where his bodily sensations, from the want of the necessary pressure, are most unpleasant: every step requires great effort, his breathing is deranged, the smaller blood-vessels begin to give way, and altogether he finds himself under conditions not suited to him. His sensations, therefore, teach him that he could only continue to exist beneath a given amount of atmospheric pressure, and that the component parts of his body are suited to it.

When we consider the effect of pressure on marine creatures, we must not forget that, from the very great difference between the elasticity of air and water, the change of relative specific gravity of an animal, rising to a given height in the atmosphere, would be far greater than that of a marine animal, unprovided with swimming-bladders or similar organs, descending to an equal depth in the sea. As far as mere fluids are concerned, and estimating the amount of change in the relative specific gravities of the creature and of the surrounding medium, which may be sustained without inconvenience, by that which a man can endure either in rarified or condensed air, we may infer that the mere difference in the relative specific gravities of sea-water and of the fluids circulating in marine animals would not be productive of very great effects even from considerable variations in depth. It is otherwise, however, with differences in pressure. A creature living at the depth of 100 feet would sustain a pressure, including that of the atmosphere, of about 60 pounds on the square inch, while one at 4000 feet, a depth by no means considerable, would be exposed to a pressure of about 1830 pounds upon the square inch. It will be obvious, from our knowledge of the structure of animals generally, that we cannot with any reference to such structure, more particularly that of the delicate vessels, suppose any one creature capable of sustaining such enormous differences of pressure with impunity.

We may fairly infer that as the sea varies in its pressure and temperature, and probably also in its amount of disseminated air, in proportion to its depth, marine creatures are formed for the conditions under which they exist; and as the latter vary, so do also the former, viewing the subject on the large scale. There is another important element to be taken into these considerations, and that element is light. How far all marine life may require light it is difficult to say; some creatures inhabiting the mud and sandbanks beneath the sea can, at all events, live long without it, and do so from choice: it must, however, be essential to all those provided with the organs of vision. Hence those which are so provided would prefer the levels where they can obtain the degree of light best suited to them: so that we might expect to find the great mass of fish, crustacea, and such molluscous creatures as possess eyes, in comparatively shallow water. When not in shallow water, and consequently where there would be less light, we should anticipate some modification of the organs of vision, by which they may be enabled in some degree to obviate the inconvenience of living in comparative darkness. Now this is precisely what we do find, and is well illustrated by the *Pomatomus Telescopium*, fished up at considerable depths off the coast near Nice, whose eyes are remarkable for their magnitude, and formed to obtain the advantage of every ray of light that can penetrate into its abode.

Our knowledge of the depths at which the animals inhabiting various shells exist is necessarily limited. Shells commonly found on sea-beaches have been thrown there by the action of the breakers; they have been moved on shore by the propelling power of the waves in that direction. Under ordinary circumstances the shells alone are ejected from the sea; it is only after heavy gales, producing great movement in the more shallow bottoms round coasts, that the animals are detected in them when cast on shore. It would appear that the motion of the disturbed water in heavy gales swept up the sands or mud in which some of the molluscous creatures habitually live, and that, being unable to resist the propelling action of the water, they are thus forced on the land.

It will be obvious that, as the action of waves sufficient to drive bodies, such as shells, on land can only be felt at moderate depths, the shells so discovered could only have been derived from similar depths. Therefore as we are not likely, under ordinary circumstances, to find deep-water shells cast on shore, it can only be by fishing, dredging, or similar means, that we obtain any knowledge of those whose inhabitants keep from choice to depths beyond the propelling action of waves. Consequently the depths at which the inhabitants of shells may exist is unknown to us; and we feel assured that there may be numerous species, and perhaps some genera, with which we cannot expect in the ordinary course of things to become acquainted. It may, however, be inferred that great pressure, and other circumstances noticed above, would limit their existence to moderate depths. Conchological works are, in general, exceedingly deficient in information as to the depths at which various shells have been discovered with their animals alive; a circumstance to be much regretted by geologists, as it deprives them, in a great measure, of the assistance they would otherwise derive from organic remains in estimating the probable depth at which a given fossiliferous deposit may have been formed. Under these circumstances my friend Mr. Broderip was kind enough at my request to construct the table placed at the end of this volume, which shows the situations and depths at which the known genera of recent marine and estuary shells have been observed.

It would appear that all the molluscs enumerated in the table noticed above have been obtained at less depths than 100 fathoms. It does not follow that many species, even of the genera noticed, may not live at greater depths than 600 feet; most probably they do so: but it is remarkable that, as far as the evidence goes, all the molluscs known to us, with perhaps a few rare exceptions, have been found with living animals in them at less depths. The area within the line of 100 fathoms, extending round the British Islands*, presents a large extent of surface, in which the bottom is at depths much beyond those noticed in the list. There may, therefore, be within what are termed soundings round our own coasts numerous species as yet utterly unknown to us. The chances are necessarily few which could bring to light such species as may habitually live in the area comprised within the lines of 60 fathoms and 100 fathoms. So long as the animals were alive, they could readily retain their places at such depths; for there can be no current or stream of tide possessing sufficient force to wash them onwards, and the action of the waves, if felt, must be very trifling. The pressure between the two lines would vary from about 180 to about 285 pounds upon the square inch, one we can easily suppose not too great for a large amount of molluscous creatures, since fish are known to live beneath a much greater pressure. We should, however, not forget that molluscs, when their organs are arranged for a pressure equal to about 200 pounds on the square inch, would probably not feel disposed to ascend to higher comparative levels where it would be much less.

We might conclude, *à priori*, that the same kinds of marine creatures are not likely to live generally under very different temperatures, or at very different depths. Now this conclusion agrees with known facts,

* See plate ii. p. 190.

and as it so closely corresponds with such facts, we may further infer that equal temperatures will not suffice for the existence of a given marine creature if there be great inequalities of depth; that is, an animal living in shallow water in the colder regions of the globe is not likely to be discovered in water of equal temperature beneath considerable depths in the tropics. And, conversely, marine creatures existing at considerable depths in the tropics are not likely to be the inhabitants of shallow water in the colder regions of the earth. Thus, when we consider the distribution of marine creatures over the globe, our views are not likely to be vitiated by the concealment, from great depths, of species known to us as inhabitants of shallow seas in temperate climates.

As we observe that marine animal life is so much affected by temperature, depth of water, and the dissemination of air, we might conclude that, at given depths in the sea and under similar latitudes, all other things being equal, we should find the same species. This is, however, at variance with all our knowledge on the subject, which seems to point to a great difference in species, with a few exceptions, under such conditions*. Assuming, therefore, that species are distinct creations, and not modifications of genera in consequence of time and place, we come to the conclusion that equal conditions do not afford like species, and that the latter have been separately created

* It was considered unnecessary to sketch the known geographical distribution of marine life, as ample information can be obtained by references to various works on the subject; and it is the less necessary to the geological reader as Mr. Lyell has treated this subject at considerable length in the second volume of his Principles of Geology.

as the places they inhabit were fitted for their existence.

This non-agreement of marine species under equal circumstances as regards temperature and depth of water, shows us how extremely cautious we should be, when viewing the subject with reference to the existing state of things, in concluding, *à priori*, that rocks are of different ages because the same marine organic remains are not detected in them; for if we are not to expect similar general collections of organic bodies under the conditions we should consider most favourable, we cannot anticipate that unfavourable circumstances will produce them.

When we turn from marine to terrestrial life, we still find that equal conditions as to climate and elevation in the atmosphere do not generally present us with like species, whether of animals or plants *. There is always a beautiful adaptation of both to the circumstances under which they are placed, but these circumstances do not necessarily afford like species. There can be no doubt that many plants can adapt themselves to altered conditions, as is readily observed in our gardens; and many animals accommodate themselves to different climates, as is well seen in those removed from temperate regions to the tropics, or from the latter to the former. But when we view the subject generally, and allow full importance to numerous exceptions, terrestrial plants and animals seem intended to fill the situations they occupy as

* We must refer, as before, to the various works which treat on the distribution of animal and vegetable life on the surface of the globe for the evidence on this head. To have entered on the subject in a volume of this kind would have too much increased its size.

these were fitted for them. They appear created as the conditions arose, the latter not causing a modification in previously existing forms productive of new species.

If the mass of vegetable and animal life were now such as it always has been, existing species, when the surface of dry land underwent different changes, entering upon the area which had suffered alteration, must have come from other situations either under similar or different conditions. Should they come from the latter, they would be capable of existing under different circumstances, and if from the former, the situations must have been contiguous, or the plants or animals must have passed over the spaces not fitted for them.

We should expect if one mass of animal and vegetable life supplied all the variations produced on the surface of the earth during the lapse of ages, that eventually there would be much uniformity in its distribution, and that fossiliferous rocks of all ages would present, taken as a whole, considerable similarity in their organic contents. Now as we do not find either this uniformity in the distribution of actual animal and vegetable life over the surface of the globe, or in the organic contents of rocks, we seem compelled to conclude either that an original mass of animal and vegetable life has been enabled to suit itself to all the changes to which the surface of our planet has been subjected since such life was called into existence, or that there have been successive creations as new conditions have arisen, so that every place capable of sustaining life has been filled by that fitted for it. There are likely to be few seeing the beauty of design manifest in creation and so apparent in animals and vegetables, who will not rather consider there has been a succession of creations as new conditions arose, than that there should be an accommodating property in organic existence which might ultimately convert a polypus into a man*.

* It is by no means intended to deny that the forms of species may not be greatly modified by the circumstances under which they are placed, for we know that such is the case. Many molluscs, animals of great importance to geologists, have been shown by Mr. Gray to alter the thickness of their shells according to the mere circumstance of being exposed to either agitated or still water. He states that "the shells *Buccinum undatum* and *B. striatum* of Pennant have no other difference than that the one has been formed in rough water, and is consequently thick, solid, and heavy; and the other in the still waters of harbours, where it becomes light, smooth, and often coloured." (Phil. Trans., 1833, p. 784.) The same author also remarks that "shells which have branching or expanded varices, like the *Murices*, are also much influenced by these circumstances, and hence many mere varieties, arising from local causes, have been considered as distinct species. The *Murex angulifer* is merely a *Murex ramosus* with simple varices; and *Murex erinaceus, M. torosus, M. subcarinatus, M. cinguliferus, M. Tarentinus*, and *M. polygonus*, are all varieties of one species. *Murex Magellanicus*, when found in smooth water, is covered with large acute foliaceous expansions; but the same shell living in rough seas is without any such expansions, and only cancellately ribbed. In such situations it seldom grows to a large size; but when it does so, it becomes very solid, and loses almost all appearance of cancellation." (*Ib.*) These and other modifications of shells, from the influence of the circumstances under which the animals have been placed, are of very great importance to geologists, as it is necessarily with the forms of fossil shells alone that they can become acquainted. And, as if still further to embarrass their judgement, it appears, though probably instances of this kind are rare, that different animals may inhabit similar shells. Thus, according to Mr. Gray, "the shells of *Patella* and *Lottia* do not in the least differ in external form, and yet their animals belong to very different orders."

CHAPTER XII.

If animal and vegetable life be so distributed over the superficies of our planet that equal conditions do not necessarily afford the same species, it follows that the organic exuviæ which may be entombed by rocks now forming, would not necessarily be the same even under the same parallels of latitude; assuming, for the sake of the argument, that the conditions as to climate, pressure of the air or water, and other circumstances, are similar under such parallels. The distribution of life, whether animal or vegetable, being thus variable as regards species, no geologist expects to find a modern rock characterized by the same organic remains in distant places, except under very favourable circumstances. He would not be surprised should he not discover in rocks now forming on the shores of Great Britain and of India a single species which should be common to both. Even when the same latitudes were concerned, he would not expect to meet with only the same fossils in the modern rocks of the coasts of Africa and of America, or of the latter and Australia; still less would he anticipate the discovery of only the same plants and animals in the lacustrine deposits now taking place in these countries, however similar the climate.

When we regard the hydrography of the world, we are struck with the fringes, as it were, of soundings round continents. These, no doubt, are of very irregular extent, the outer verge, or the fall from about 150 or 200 fathoms into deep water, sometimes approaching the land, at others receding from it, the differences being generally due to combinations of various local circumstances. The general character, however, of these fringes, is remarkably similar. They constitute plains, for the most part not much inclined, to the depth of between 600 or 1200 feet, where there is very frequently a more sudden plunge into much deeper water.

Fig. 43.

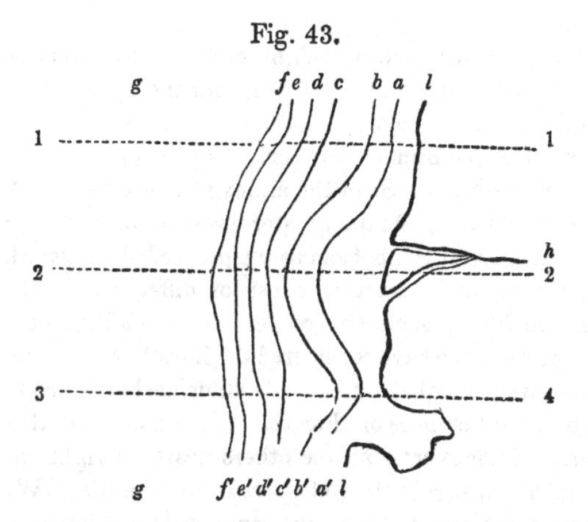

Let, in the annexed wood-cut, (fig. 43.) 1, 1; 2, 2; 3, 3, represent parallels of latitude sufficiently distant from each other to produce considerable variations of temperature, *l l* a line of coast extending from north to south, and *f f* the outer verge of soundings off the same coast, there being a sudden fall into deep water *g g*. Let the lines *a a'*, *b b'*, *c c'*, *d d'*, and *e e'*, represent lines of equal depths between the shore and the verge of soundings. It is obvious that if we consider the soundings to form an inclined plane from the shore to the verge, which may, for the sake of illustration, be supposed to run out to a depth of 150 fathoms, we should have a series of what may be termed zones of different depths, and, consequently, of different temperatures and pressures. We might infer that each of these zones would be inhabited by creatures which should on the whole differ from each other.

Neighbouring zones might contain some marine animals common to two or three, but the resemblance in this respect would become less the further the zones were removed from each other. Supposing the coast line to be N. and S. in the annexed figure (fig. 43.), we should then obtain greater pressure and a lower temperature at the bottom as we proceeded eastward. There would be another cause of difference in the marine life of such an area as the soundings here supposed. We have considered the lines 1, 2, 3, as representing parallels of latitude sufficiently distant to produce differences of climate. We then have another series of zones crossing the others nearly at right angles, and also affording changes of marine life. We should not expect to find the same creatures between 1 and 2, and 2 and 3. We thus obtain various areas, which not only differ from each other in east and west directions, but also from north to south; and any accumulation of detrital or other matter, forming new rocks over the whole area represented, would entomb remains differing from each other and yet be contemporaneous. We have also introduced in the above diagram (fig. 43.) another cause of change. The river *h*, discharging itself by a delta into the sea, would cause considerable modification in the life within the influence of the fresh water propelled outwards. The organic remains brought down by the river would produce a general modification in the zoological and botanical characters of the new-forming rocks, and there would be a mixture of fresh-water and marine creatures, and probably also of terrestrial animals and vegetables, over a given part of the soundings.

We have assumed, for the sake of the argument, that the soundings constituted an inclined plane; this would not be strictly the case in nature, for, generally speaking, there are minor elevations and depressions forming banks and hollows, depending on local circumstances. We shall, however, continue to consider these beds of soundings, fringing the land, as inclined planes, for the sake of more easy illustration. Let *a b*, in the annexed diagram (fig. 44.), represent the

Fig. 44.

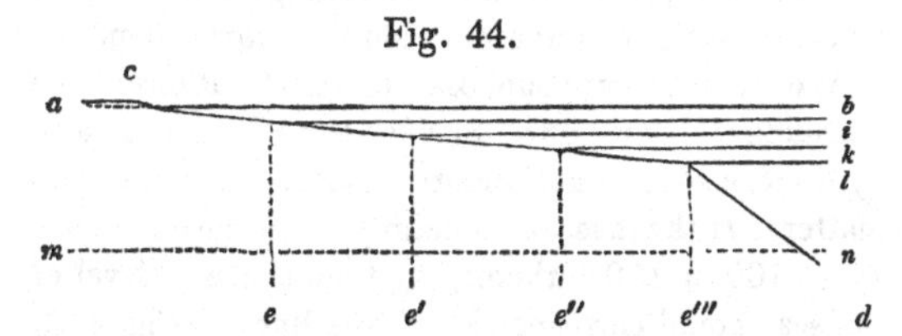

sea-level, *c* a coast, and *e*, *e'*, *e''*, *e'''* a section of a mass of soundings, falling into deep water at *d*. At the various depths, *i*, *k*, *l*, there would be different temperatures and pressures; and, from what has been previously stated, the respective portions of the inclined plane of soundings, cut by the levels of depth, would be tenanted, viewing the subject generally, by different marine creatures.

If, now, the whole body of land, soundings included, become gradually raised, so that the level *i* forms the surface of the sea, the creatures which lived formerly between the surface *b* and the level *i* could retreat to the portion of bottom situated between *i* and *k*, where they would be under the same circumstances as before. There would, however, be one result, which, when viewed geologically, possesses considerable interest; the organic remains entombed by an accumulation of detrital or other matter on the surface comprised between $e' \, e''$ would no longer be the same, considering the subject generally, as they had previously been. A change will be effected. The exuviæ, which in the supposed original state of things would be entombed in the surface $e' \, e''$, would now be covered by others corresponding with those previously enveloped by new-forming rocks on the surface $e \, e'$. The like would happen with the succeeding surface $e'' \, e'''$, whose existing inhabitants would seek lower depths.

We have hitherto supposed minor elevations of the soundings; let us now consider the effects produced by a more sudden and greater rise of the same solid matter. If the mass of soundings were raised at once about 100 or 150 fathoms, so that the new level of the sea should correspond with the line *m n* (fig. 44.), not so suddenly as to cause great waves, but sufficiently quick to prevent the migration of the molluscous and other inhabitants of the bottom, these creatures would necessarily be destroyed as far as regards the particular area. Those which were incapable of quick locomotion, as would be the case with many, would perish if the elevation were moderately gradual: they would be unable to reach the new lines of depths and temperatures suited to them.

Let the reader consider what would be the effect upon marine animal life (at least that part of it which, unlike the fish and some others, cannot readily change their places,) if the area of soundings round the British Islands were so raised that the present line of 100 fathoms in depth constituted that of the new coast*. To those unacquainted with geological phænomena such a supposed change of level may appear somewhat great; but it is one which such phænomena show us, in a manner not to be mistaken, has often occurred on the surface of the earth, and that so far from being great, it is comparatively small. To the geologist such a change of level presents nothing remarkable; the only question likely to arise is simply whether such changes may have been gradual or more or less sudden. In either case we should have changes in the condition of the marine life at present existing on the area. If the elevation were moderately sudden, so that the various creatures habitually living on or in the bottom of the sea should not have the power to escape as a whole, there would be an immense destruction of species, the greater part of those peculiar to the area probably perishing altogether. Numerous shallow basins containing sea-water would remain, and in these many creatures would continue to exist for some time longer.

Those who have seen the tide leave the shore on

* For this area see the map opposite p. 190.

shallow coasts, and witnessed the rapidity with which large tracts of slightly inclined sands or mud become freed from water, pools here and there appearing, will readily conceive that a rise of bottom, producing as rapid a disappearance of water, would entirely prevent the great mass of molluscous creatures, and even some fish, from escaping. Yet the disappearance of the water would be so gradual that the surface would no more be torn away, than is the case when the tide ebbs on shallow coasts. In all probability a much quicker disappearance of water might take place and yet the bottom remain firm. The rapid disappearance of water by an ebb-tide is, perhaps, less familiar to most persons than its quick flow on similar coasts, because it is unattended by the same amount of danger. If, therefore, they measure the rapidity of the relative change of levels by that of the rise of flood-tides on shallow shores, they would better perceive how difficult it would be for a large amount of marine creatures to quit their usual habitats and escape; for this rapidity is on some coasts so considerable that men have great difficulty in reaching the shore, though conscious of the danger, and in some situations it even requires the aid of a good horse to escape in sufficient time. Now it is important to remark, that the marine animals, being unprepared for such a relative change of level, would scarcely be aware of their danger until too late. Those habituated to great tidal changes would of course perish at once; for when left by the water they would not feel uneasy, and therefore desirous of change of place, until the time when the tide usually returned.

It is evident that if a change of level took place in

this part of the earth's surface, even accompanied by no greater rapidity in the disappearance of the sea than that observable in an ebb-tide, so that a new line of coast should correspond with the present line of 100 fathoms in depth round the British Islands, there would be great destruction of marine life, more particularly of such species as were peculiar to the area left dry, and not possessed of quick locomotive powers. Even if many escaped to the outer verge of the present line of soundings, the area, from the greater steepness of the bottom (the descent into rapid water being comparatively rapid), would be so narrow that, even supposing all other circumstances fitted for their existence, very few, even of those capable of speedy flight, would eventually find places proper for them.

Let us now consider another consequence of this change of level round the British Islands. In many situations shallow basins would exist into which some creatures might escape. These basins would, however, have been previously tenanted by other marine animals, which, if the change of level were inconsiderable, would be capable of living in these basins, at least for a time. If, however, they were creatures which required comparatively deep water, they would probably perish, and the waters be inhabited by those which may have escaped from higher levels. These basins would not long exist under their first conditions. The new surface of the elevated area would come within the action of the atmosphere, and be subjected to all the effects produced by such action. The rivers of a part of France, of all the British Islands, and of those parts of Germany, Russia, Sweden and Norway, the drainage of which now runs into the area under consideration, must proceed onwards until they reached the sea. They must find their way through various parts of the new dry land. At first, from the difficulty of finding channels, there might be considerable spread of water and consequent loss by evaporation; but in a climate like that under which this change would be effected, the accumulated waters would soon cut channels, and meander to the sea by various routes, each river according to the least resistances opposed to its progress.

A difference of 600 feet in level over the area would no doubt tend to throw some of the waters of continental Europe backwards, supposing the rise somewhat local and to fine off to different distances around it, and some of them would take other lines of drainage; but much of the water would still flow westward, and the drainage of the area itself would not be inconsiderable. Many of the basins above noticed would receive rivers in their courses onwards to the sea, and we might expect from this constant influx of fresh water that the sea water would gradually become less saline, the waters mixing and the mixture passing off over the lower drainage lip of the basin, for a drainage must take place when the mixed waters rose above the level of its sides. Many of these basins would become freshwater lakes, and consequently would be ultimately tenanted by freshwater animals brought down by the rivers; for this change of level would leave terrestrial and freshwater creatures in a great measure unharmed. They might, indeed, shift their habitats to lower comparative levels, but destruction would not ensue: they would be free to colonize the new dry land, and therefore their remains would be entombed in any deposits forming in the lakes, which, from all analogy, we should expect to happen.

It is somewhat curious to consider the zoological character of the beds of sand, silt, or mud which would be left dry. Charts of the various soundings show us that we should have a surface of variable mineralogical character, but that it would be principally arenaceous. Now, as among marine creatures some prefer one kind of bottom, and some another, we should not only have zones of marine life, if we may use the expression, corresponding in a general manner with those of equal pressures, temperatures, and other circumstances, but such marine life would also differ, according to the nature of the bottom in the zones themselves. We should thus have an area in which contemporaneous creatures were entombed, where the deposits in one portion would differ in organic contents from those of another, in proportion to differences of depth, temperature and bottom. The lakes would eventually contain the exuviæ of terrestrial, fluviatile and lacustrine life, both animal and vegetable, and we should have rocks characterized by such remains resting on others full of marine exuviæ, with probably here and there some intermediate beds containing remains characteristic of brackish water, or of a mixture of creatures, originally either marine or fresh-water, which had been enabled to support the changes of the medium in which they occurred up to a certain point, when the originally marine creatures gradually disappeared and gave place to the fluviatile and lacustrine animals.

We have observed that the soundings round the British Islands are bounded on the side of the ocean by a line, where the great plain, as we may term it, terminates somewhat abruptly, the line of 200 fathoms being at no great comparative distance beyond that of 100 fathoms. Now, if the change of level were such that the new coast line rose somewhat suddenly, even for a few feet above the ocean-level, the breakers would attack it, gradually removing the land, and producing cliffs, which, though not at first high, would become more so as the inroads of the sea increased. If, however, on the contrary, the new coast did not rise abruptly above the sea, but was low and bounded seaward by a zone of shallow water, the piling influence of the sea would produce dunes, similar to those on the low shores of western Europe extending from France to Denmark, and to those which separate the flat district of the Landes from the Bay of Biscay. Behind such dunes we should have masses of pent-up waters, such as are commonly found in similar situations, and consequently similar results as to the kind of animal and vegetable life entombed, and to the nature of the deposits formed.

It will be obvious that the less the amount of relative change of level produced at any one time in this area, the less would be the general amount of destructive effects caused by it; so that if the change be brought about by no greater exertion of force than that supposed to be now observable in the Gulf of Bothnia, there need be no destruction of species, the inhabitants of the shallow waters gradually retreating with the sea, so that there would be a general move of marine creatures seaward, each particular species keeping the same relative positions as regards pressure of water, temperature, and other necessary circumstances, as before. One remarkable geological effect would, however, in this case be produced, and will appear more clearly, if we suppose the retreat of the sea to have been thus gradual to the verge of the soundings, so that the line of 100 fathoms constitutes the coast. The whole of the surface would, to a very great extent, be covered by the remains of similar species, left on the bottom and enveloped by the usual causes as the sea retreated, and such species would be littoral. The deposit would clearly not be contemporaneous, but it would be characterized by similar organic remains. It would also be based on other deposits formed in a similar way, but characterized by the presence of organic remains differing from the former, because they are the exuviæ of creatures which resided at greater depths; they would, however, be similar as regards themselves.

Fig. 45.

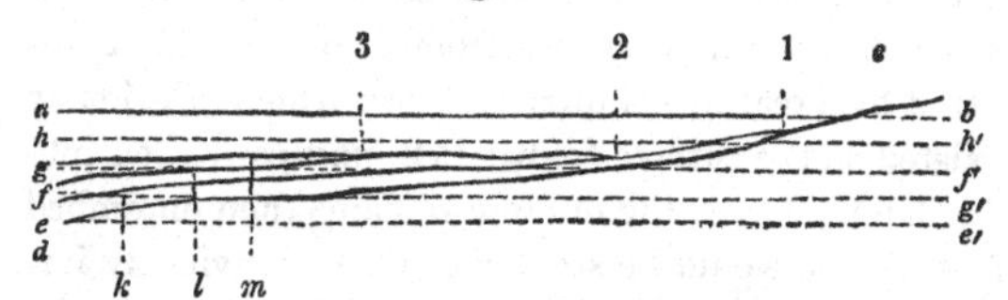

Let *a b*, in the above section, fig. 45., (and unfortunately we are compelled to make it exceedingly exaggerated on account of the small size of the page,) represent the level of the sea; *c*, the land rising above it; and *c*, *d*, some surface of rock upon which detritus or other matter forming beds, has accumulated. Let us assume that, from the ordinary operations of nature, a deposit takes place beneath the sea on the surface *c, d*, and that this deposit constitutes a bed conformable to *c, d*, extending from 1, to and beyond *d*. The animals inhabiting the bottom, and whose exuviæ, from the death of the creatures or other causes, are discovered in such a state as to lead us to suppose they had not suffered transport, but were entombed in the sands or mud, in which they lived, would probably differ in a general manner, according to the depths of water above them. They would be suited to the various levels *h h'*, *g g'*, *f f'*, *e e'*, and consequently the organic remains are not likely to be precisely the same at the depth *d*, and near the coast *c*, though entombed in a bed of contemporaneous production.

If, now, the sea retire from 1 to 2, so gradually that the species are not destroyed, but as gradually change their places, retreating seaward, there would be a general move at the various depths in the endeavour of the animals to keep under conditions, as to pressure and other circumstances, similar to those under which they lived prior to the relative change of level. Consequently, the new surface from 2 to 3 would be covered by creatures similar to those which previously existed on the surface from 1 to 2, now converted into dry land. Hence, the organic remains then entombed from 2 to 3 would be similar to those previously imbedded from 1 to 2. The like would happen if the surface from 2 to 3 became dry land, the littoral creatures moving to another lower surface, and so on to the various depths. We may thus obtain the successive beds *k, l, m*, which, though they present similar organic remains horizontally, are not contemporaneous. Unfcrtunately the diagram (fig. 45.) is necessarily so exaggerated, in order to render it at all intelligible, that the close approach of the beds to horizontality, and consequently the extremely small angles at which the lines of equal depths would cut the consecutively formed beds, is not apparent, and may require some little trouble to conceive. When we consider the area zoologically, it will readily be admitted that it is variously inhabited according to temperature, depth of water, and bottom, and that the influence of the two former tends to produce zones of different kinds of animal life. When, however, attention is called to the angle which the surface of the bottom of the area under consideration, viewed generally, forms with the horizon, there is great difficulty in showing those who couple ideas of vast depths with all seas, that this angle is as small as it really is. It probably does not amount to more than that which the flats, as they are termed, uncovered to a great extent by low water on certain coasts, make with the sea at the same time; if indeed it be so great. It is necessary to call attention to this circumstance, otherwise it might be supposed that the beds noticed in the diagram would present such obvious changes in their zoological contents that no one could possibly avoid perceiving the difference of organic remains in the contemporaneous deposits, and consequently not imagine that the same kind of exuviæ was characteristic of only one deposit and one epoch. When, however, the angles above noticed are taken into consideration, the readiness with which the two things may be confounded will become apparent*.

* The author has purposely avoided all consideration of the area being part of the surface of a sphere or spheroid, and con-

It will be obvious that if instead of the change of levels here supposed, and which should convert a large area into dry land, there had been a subsidence of the bottom of the same area to the amount of 100 fathoms, by which the previous line of 100 fathoms became covered by 200 fathoms of sea, and so in proportion over the surface, we should have effects the reverse of those we have previously noticed. As we supposed the greatest elevation to be beneath the British Islands, so we must now consider the greatest subsidence to have been in the same place, the amount of depression becoming less as we receded from these islands as a centre. If the subsidence were gradual, there would appear no reason, from this cause alone, why any species should become extinct, as they would not be left dry, and could readily work their way up to the levels proper for them. A large portion of the British Islands would be submerged, as would also be the case with a considerable part of Western Europe, even considering the depression not to fine off at a very small angle. Great physical changes would necessarily result, more particularly from the action of the waves and streams of tide among those parts of the British Islands which rose above the surface of the water in the form of numerous isles. There would be a tendency to equalize the depressions by the deposit of detritus, and consequently the remains of organic life, both animal and vegetable, would be entombed to a considerable extent.

sequently that the lines are curves and not straight, as it did not appear necessary to the illustration of the subject under discussion, and would only tend to render it still more obscure to the general reader.

Numerous fluviatile, lacustrine, and estuary deposits would, under these circumstances, be covered by marine formations, gradually or suddenly, as the change of levels had been either slow or rapid; in the first case we should often have transitions of one kind of life into the other, while in the latter the change from one deposit to the other would be somewhat abrupt. On the bottom of the previously existing sea there would be a general move of marine life towards the new lines of shore, in order to accommodate itself to the new circumstances under which it should be placed, and thus the remains of animals habitually living in the deeper water might cover those of creatures whose habitats were in minor depths, and which had been deposited when the waters were generally less deep.

Although marine species would not be affected by this change of level, it would be otherwise with terrestrial life, whether animal or vegetable, and in certain cases, with the inhabitants of fresh waters. As the sea rose relatively to the land, the lower lines of vegetation, if not washed off the surface and carried away by the action of the waves, tides and currents, would be imbedded to a certain extent in the rocks formed round the shores. A comparatively sudden rise of water would greatly destroy the lower zones of vegetation, though probably seeds would escape and be thrown on shore, so that eventually there would be much the same kind of vegetation round the smaller isles, as at present exists in the lower parts of the British Islands. As, however, these islands would be reduced throughout by 600 feet in height, we should anticipate some change in the vegetation of the higher regions, more particularly when we consider the alteration of climate which, to a certain extent, would ensue from differences in the relative amount of sea and land in Western Europe, and of the height of the land above the ocean-level. Terrestrial animal life would also have to accommodate itself to the new circumstances under which it would be placed, and much of it would necessarily be destroyed, though the species might not become locally extinct, at least not to any great extent. As great tracts of marsh occur at the lower levels, they would be the first covered by the sea, and their inhabitants must change their quarters to escape, not only from the inroad of the sea itself, but from the voracious creatures which such an inroad would bring with it. They would with difficulty find situations fitted for them, unless the relative change of level was so gradual that sufficient areas of marsh could be prepared for them in proper time, which, considering the present rate of increase in such situations, would scarcely be probable. At all events the area of dry land would be much diminished, and the same amount of animals could not find room in it: there would be considerable collision of species against species, abstracting man entirely from this imaginary condition of things, intended to illustrate geological events which occurred anterior to his existence on the earth. The weaker would give way, and thus some species might be exterminated so far as the islands were concerned.

We have thus dwelt at considerable length on the consequences of relative changes in the levels of sea and land, to the amount and in the situation above noticed, because they are such as will assist us in explaining many geological phænomena, particularly among the fossiliferous rocks. We there find changes in the zoological or botanical characters of deposits, sometimes abrupt, at others exceedingly gradual, with very frequently no trace of a wash of waters over the lower rock. There are no doubt many situations where there has been considerable abrasion of a lower fossiliferous rock before another fossiliferous rock was deposited upon it, but the localities are exceedingly numerous, where, notwithstanding a complete, or nearly complete, change in the nature of the organic remains, there is nothing that marks an abrasion of the lower rock prior to the deposition of the beds which repose upon it.

It is stated that certain species of shells are common to the shores of Western Europe and to the eastern coast of North America, from whence it would follow that deposits now forming on the opposite shores of parts of the two continents may contain some of the same species of shells. In this case contemporaneous deposits, though at considerable distances, would afford some of the same organic remains. Such deposits are obviously not continuous: they merely form, as it were, a part of the fringe of soundings on the shores of both continents; for the two coasts are separated by great depths of water, at the bottom of which the inhabitants of the shells in question could not exist. We learn by this fact that equal circumstances as to pressure of the surrounding water, temperature, light, food, and power of procuring air, may exist in two different localities, and the eggs of marine animals being transported by natural causes from the one to the other, deposits in the progress of formation in such localities may, though not continuous, contain somewhat similar organic remains. The same effects would, under similar circumstances, more readily take place at

Fig. 46.

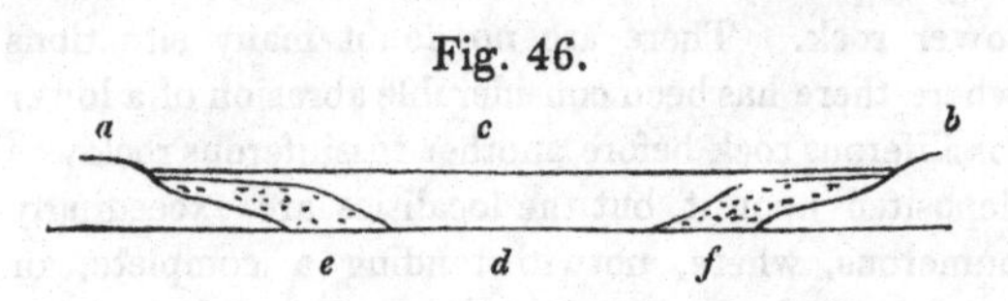

minor distances; and the less the distance the greater the probability, under the equal conditions above noticed, that the zoological character of the two deposits would be alike. Let *a* and *b* (fig. 46.) be two opposite coasts under equal or nearly equal climates, and *e*, *f*, two rocks, or accumulations of sand, silt, or mud, now increasing under similar conditions, and enveloping the remains of molluscs living in and upon their surfaces. Now if the eggs of molluscs can be transported across the sea, *c*, from the one shore to the other, there will be a tendency to equalize the relative kinds of creatures in each mass of new-forming rocks, *e*, *f*, in proportion to the proximity of the two coasts *a*, *b*, all other things being equal. The deep water *d* presents a barrier, not only to the progress of the molluscs by the bottom of the sea, but to the transport of the detritus itself, and therefore we have deposits evidently of the same age characterized by similar organic remains, without ever having been continuous; a circumstance which should be borne in mind when we assume the former continuity of rocks solely because their zoological and botanical characters are similar.

We have already shown that under the supposition of a rise of the bottom of the sea round the British Islands, to no greater amount than 600 feet, we might have deposits, if the relative change of level were gradual, formed at different times and characterized by the presence of similar organic contents. We will now view the subject in a different manner. If two coasts were situated relatively to each other as represented above, and under the same conditions, (fig. 46.) and the one were subjected to oscillations, rising or falling beneath the general ocean-level, while the other remained firm, we should obtain a series of results in the one case different from those in the other. While the one contained exuviæ characteristic of a marine deposit, there might be great variety, not only in the kind of life entombed in the other, but also in the mineralogical structure of the rocks formed at the same time. It might also happen that a destruction of species took place in the one situation, while they continued to exist as usual in the other. Hence we should have little or no resemblance in the organic contents of contemporaneous rocks, though if there had not been these oscillations in the one situation, while the other remained firm, the conditions as to temperature, pressure of water, light, food, and the power of procuring air, would have been such, that the organic remains would have been, to a certain degree, similar in both.

CHAPTER XIII.

We have hitherto considered marine animal life with reference to creatures living in particular depths of water, and which, when we regard those of the ocean, are comparatively shallow. The surface of the deep ocean teems with animal life, and is probably not less well proportionably tenanted down to depths where, from the want of the necessary conditions, it ceases to exist. Fish, crustacea, and molluscous creatures occur in the open ocean, and no doubt afford by their mutual destruction abundance of food to each other. To remain at ease, these various creatures must be in a state of equilibrium as regards the water which surrounds them. Most of them, if not all, have probably the power of altering their specific gravities to a certain extent, by which they are enabled to rise and descend to and from given levels. Nothing is more curious than to see the ocean in a calm suddenly covered with marine life of various kinds which must have risen from beneath the surface, where it remained until the circumstances previously noticed were suited to its presence. All these creatures have no doubt limits to their powers of either increasing or decreasing their specific gravities, so that we may readily conceive a stratum of water, if we may use the expression, of a given thickness from the surface of the ocean downwards, replete with animal life, the amount of life in the stratum itself, viewed on the large scale, decreasing with the depth, or in other words, as the various conditions for its existence become unfavourable.

What the actual depth may be beneath which animal life does not exist in the ocean, we have yet no good data for ascertaining; but that there must be such a limit, to at least any kind of life analogous to that with which we are acquainted, there can be little doubt. Now though this life may be, and probably is, principally supported by a system of universal voracity, one creature devouring another, every creature being moreover provided with the best means of obtaining its prey, we can scarcely conceive this kind of reciprocal and universal banquet so complete that some creatures may not here and there escape and die natural deaths. When a creature did come to its end naturally, the chances of its not being devoured would still be small, for at the time of its death it either was or was not in equilibrium as to the surrounding aqueous medium. If it were at a depth which it required the power of muscular exertion to retain, death would terminate this power, and the creature would rise or fall as the case might be, like any inorganic body, and arrange itself according to its relative specific gravity. It would only descend to the bottom of the ocean if it were specifically heavier than the waters of the sea.

Now though numerous creatures living in the open ocean contain parts of much greater specific gravity than sea-water, their bodies, including fleshy matter, are, taken as a whole, generally of the same density as sea-waters at moderate depths. Indeed this seems almost a necessary condition for their habits. When a marine creature dies, therefore, it would probably remain suspended in the water at levels not very different from that in which it parted with life, unless the gaseous contents of air-bladders expanded considerably from the want of muscular compression. Decomposition of the fleshy matter would, among other things, produce gaseous compounds, and hence the relative specific gravity of the body being diminished it would rise towards the surface. We are, however, not to suppose that the marine scavengers, of which there appear to be many, would be idle; and there is little probability that the body would long remain undevoured even in its putrid condition. Upon the kind of creatures by which the body may be devoured, would depend the chance of its solid parts, particularly an entire part like a whole shell descending to the bottom. If, for example, the body of a Nautilus were consumed by minute crustacea, it would sink as the matter which rendered the dead creature as light as the surrounding water passed into the animals feeding upon it. Finally the shell of the Nautilus would descend to the bottom, for in all probability no part of the waters of the ocean is likely to possess a density of 2·64, which would be about that of a Nautilus, judging at least from the specific gravity of *N. umbilicatus* *. Should the dead Nautilus be crushed by the jaws or palates of a large fish, the only chance of the descent of parts of the shell to the bottom would be after it had passed through the animal.

Altogether, we should expect any solid matter, such as shells or their parts, deposited at the bottom

* See the specific gravities of several shells, p. 76.

of the deep ocean, to be differently circumstanced from those occurring in or on the bottoms of more shallow seas. In the latter case multitudes of thin delicate shells may readily be found entire, more particularly in the mud and sands, which for the most part constitute their retreats. When a thin-shelled mollusc dies in sand or mud, its body is already entombed, and the forms of the shells are such as successfully to resist moderate pressure for some time, during which there may be an infiltration of foreign matter, producing an equal if not greater power of resistance than was afforded by the original fleshy body of the mollusc.

When we attentively study the mode in which animal organic remains occur, we perceive that, though there are very considerable differences in the amount of their preservation, a very large proportion of them must have been entombed uninjured, and many alive, or, if not alive, at least before decomposition ensued. Skeletons of Ichthyosauri and Plesiosauri are sometimes found in the lias, with their various parts so connected, (traces of skin being even observable, and the remains of indigested food still between their ribs, and where the stomach must have existed,) that we can entertain little doubt they have been entombed with their flesh upon them in the mud, frequently calcareous, from which the lias originated. Shells are often distributed in a bed in groups precisely as they would be at the bottom of the sea, and with their parts so perfect that they appear merely covered up by matter which has accumulated upon them. At other times whole beds are composed of broken shells, showing that they have been exposed to sufficient violence to crush them, before they were accumulated in strata. The arming of the sounding-lead so frequently brings up little masses of broken shells from the bottom, that there can be little doubt that beds of broken shells exist there, and only require to be covered with sand, mud, or calcareous matter, and then elevated above the level of the sea, to afford us strata placed similarly to those we see in many rocks.

We may fairly infer that the great mass of organic remains, like the accumulations of detritus among which such remains occur, was accumulated in moderate depths around coasts, and either on or around shoals at no very considerable depth beneath the surface of the sea. The chance of finding the greater part of Saurian and other remains of that class, and of terrestrial creatures near the land, is obvious. The latter creatures would have, as has elsewhere been observed*, but a poor chance of escape in a sea amid voracious animals, all, great and small, ready to devour them. The greater part of fossil Saurians may be inferred from their structure to have lived near, or partially on, land. How far the Ichthyosaurus may have been enabled to keep the open sea, we know not, but its structure seems as well adapted for the purpose as that of a porpess.

We refer beds to a freshwater, estuary, or marine origin according to the organic remains found in them, and which we may consider analogous to forms now discovered either in rivers, lakes, estuaries, or the sea, or introduced into them from the land. This of

* Geological Manual, p. 346.

course assumes that particular forms have been constant from the earliest existence of animal and vegetable life, either to the waters of these respective situations or to the land, and that the forms of vegetables or animals of which there are no living analogues are such as to warrant our classifying them correctly either as marine, freshwater, or terrestrial. Generally speaking, the inferences which have been made as to the probable former habitats of the vegetables and animals, the remains of which are entombed in various rocks, seem so well grounded on analogy, that we are little disposed to doubt them. All the great and leading facts of this kind have probably been fairly explained, yet a little caution seems necessary in minor points. Should, for instance, a *Voluta* or an *Arca* be detected among the organic contents of a rock, it would at once be considered as marine, because both genera are generally known to us as marine, and probably in most cases the inference would be correct; there is, however, a chance of error in both instances, for *Voluta magnifica* is known to live high up in the brackish waters near Port Jackson in Australia*, and an *Arca* inhabits the freshwater of the Jumna, near Hamirpúr, 1000 miles from the sea†. Perhaps if both these shells were examined attentively, some modification of form would be observed: such attentive examination of the form in a fossil is, however, not always given, nor does it

* From the information of Mr. G. B. Sowerby.

† Gleanings of Science, vol. i. p. 265, Calcutta, 1829. Also from the information of Mr. G. B. Sowerby, who likewise states that a *Nucula* is discovered in the Ganges at Banda.

appear very clear, *à priori*, what kind of modification we should expect to find.

The circumstances which might compel an animal to live under conditions not common to the particular genus or species to which it may belong, should also be well weighed. Mr. G. B. Sowerby informs me that an *Astarte* and a *Cardita*, (neither of which is yet named,) two genera commonly considered marine, were found in pools in the ice near Melville Island, therefore in waters, if not fresh, not very saline. It was remarkable that the umbones of these shells were worn, as is very commonly the case with those of freshwater molluscs. I am indebted also to the same experienced conchologist for the fact that *Anodon anatinus*, a common freshwater creature, lives abundantly in the West India Docks, London, and consequently in at least brackish water. That it thrives and breeds there is also ascertained, and one individual was found covered with thousands of its eggs, showing that it flourished as much in the Docks as in pure fresh water. The same mollusc is also found in the Commercial Docks, where it is accompanied by *Mytilus polymorphus*, a Danube and Volga shell, brought perhaps originally on the bottom of some vessel from the Black Sea.

Dr. MacCulloch instituted a series of experiments, which enabled him to prove that "the turbot, sole, plaice, mullet, smelt, atherine, horse-mackarel, pollock, loach, basse, rock-fish, whiting-pout, rockling, prawn, crab, and stickleback, may be habituated without difficulty to fresh water; while all those which have hitherto had time enough, have bred and propagated" *. The same author calls attention to other circumstances of a similar kind, such as, that the cod lives from choice in a freshwater lake in Shetland, and that the herring has been discovered as a constant inhabitant of another body of fresh water. That freshwater fish, pike and *Cyprini*, live from choice in the salt waters of the Caspian, while the waters of the Volga are open to them, has been long known. The fact also that salmon and other fish of the like habits can readily support changes from salt to fresh water, and *vice versâ*, has often been noticed. It would appear from the experiments hitherto tried to cause marine creatures to live in fresh water, and freshwater animals in sea-water, that if they be suddenly taken out of one water and plunged into the other, they die; while if the water be carefully and slowly altered from one condition to the other, many creatures support the change. When freshwater lakes behind shingle- or sand-beaches, thrown up by the sea, are made to communicate suddenly with the sea either by natural or artificial means, the sea-water, from its greater specific gravity, rushes as a body to the bottom of the lakes, and destroys the freshwater fish that it may meet with. A few remarkable instances of this kind have been observed on our own coasts. Had the changes been effected gradually, very few of the freshwater creatures would probably have perished.

There is also another circumstance to which attention should be paid. Many freshwater springs exist in the sea, and will be more or less copious according to circumstances. Several of these springs have been

* System of Geology, vol. i. p. 330. 1831.

long known, and that in the Gulf of La Spezia, which discharges a very considerable quantity of water per second, has been celebrated since the days of Pliny. It rises with considerable force, so as to produce a slight convexity on the surface of the sea, and its waters are probably derived from a system of cavernous passages in the neighbouring limestone rocks. There must be many similar springs in various parts of the world which remain unknown, because the fresh water does not rise with sufficient force to the surface. There must also be thermal springs as well beneath the sea as on dry land, for the forces which cause their discharge seem sufficiently powerful to overcome the pressure of the sea. Moreover, if fresh water occur anywhere beneath the waters of the ocean, it will tend to rise from its inferior specific gravity without taking into account its temperature, which, supposing the springs thermal, would also cause their waters to rise. We should expect that such springs would produce an effect on the kind of life within their influence. The following fact may perhaps be taken in illustration of such an effect, though part of the evidence is not so clear as could be wished. Fresh water rises through the sea, from springs in the bottom beneath, on a part of the coast of Java, not far from Batavia. Now, it appears that *Cyrena zeylanica*, considered a freshwater shell, is obtained in great numbers in the sea near the same coast, and it is deserving of remark that the umbones of these shells are worn precisely as they so commonly are in those inhabiting fresh water *. It was not di-

* From the information of Mr. G. B. Sowerby, who learned the fact from Mr. Hardy, of the Madras establishment.

rectly proved that the water was fresh, or nearly so, where these shells were obtained; but the inference is highly probable that they lived amid freshwater springs. How far some individuals might be enabled, first to live in brackish water, and finally in that of the sea, so that there should eventually be two races descending from the same stock, one marine, the other freshwater, is a separate question; but from what has been above stated respecting similar changes, such an adaptation to circumstances is far from impossible.

To obtain uniformity of organic life over the face of the globe, the conditions for its existence should be similar, or nearly so. Consequently, when we suppose that given rocks are characterized by similar organic remains, we also infer similar general conditions of matter, whether aërial, terrestrial, or aqueous, over the area where such rock was formed, and in which the animals and vegetables existed prior to the deposit of the rock. We have seen that the distribution of animal and vegetable life is at present exceedingly variable, even under similar conditions as to climate and other necessary circumstances, so that actual contemporaneous and fossiliferous deposits in various parts of the earth's surface, would not contain similar organic remains.

We have next to consider what conditions, if any, could produce greater uniformity in this respect. Upon the supposition that our planet is a mass of matter which has gradually become more cool from the radiation of its original heat, we should first have climates in a great measure independent of solar heat, though not of solar light. When the surface of the earth would become sufficiently cool to permit the existence of animal and vegetable life, we also infer, from the structure of that now existing on our globe, the existence of water in its liquid state.

From all analogy, we should not suppose that water at high temperatures could support animal or vegetable life, but that its heat must be at least moderate. Now the distribution of a thermal ocean over various parts of the globe would have great influence on climate, and there would be a great tendency to equalize temperatures from this cause alone. The moist condition of the atmosphere, as it would be probably loaded with aqueous vapour under such conditions, would have little influence on marine life, except as regards light; but it would be otherwise with terrestrial life, which would necessarily be much affected by it. One principal effect from the nearly uniform temperature of the atmosphere at equal heights round our planet, might be great comparative stillness compared with its present state. The equatorial and polar regions being at equal, or nearly equal temperatures, the great cause of the present system of winds would be wanting. In fact, all those atmospheric phænomena at present due to the heat of the sun would scarcely be observable, and would only become apparent as the radiation of heat from the earth threw the atmosphere, as it were, under solar influence.

Under the conditions here supposed, the waters of the ocean would, however, obey the laws by which they are governed, and would arrange themselves according to their relative specific gravities, and consequently, according to their relative temperatures. Under the above circumstances there would necessarily be great uniformity of temperature, the only variations of importance arising from difference of depth, the cooler water being lowest; for we may readily take the temperature of the waters to be above 40° under these conditions. There might be differences in marine life arising from differences of depth, but we should have a state of things highly favourable to horizontal uniformity. As, however, the heat of the earth's surface gradually became less, these conditions must necessarily have altered, until those we now witness finally obtained. If we admit great heat within our planet, which has, during a lapse of time, (difficult, from our limited ideas on such subjects, to conceive,) gradually by its decrease permitted the surface to assume its present state, and consequently allowed the sun to exercise its present influence, we obtain a series of conditions, the lowest term of which is highly favourable to uniformity in the distribution of animal and vegetable life over the face of the globe, as far as regards the circumstances we have mentioned, while the highest term is not thus favourable, the intermediate terms presenting conditions resembling the two extremes in proportion as they respectively approached or receded from them. To what extent uniformity in this respect has prevailed, admitting the theory of central heat, is another question, and its solution will be best sought through the medium of facts; but, *à priori*, we may be led from the analogy of actual animal and vegetable life to suspect, that as similar conditions do not necessarily now afford similar kinds of life, prior states of the surface may also have presented like variations. We should, however, still expect to have far greater uniformity formerly than at present, in as much as the chances of such uniformity would be far greater in the one case than in the other.

Organic remains are presented to us under various kinds and degrees of mineralization. Some have suffered little alteration, while others show a substitution of one substance for another; so that, in fact, we have no part of the original body, but a mineral substance of exactly the same form. There can be little doubt that these various states are due to the conditions under which the organic bodies have been placed since they were entombed in rocks, both as regards time and situation.

Shells, whether terrestrial, freshwater or marine, are composed of carbonate of lime mixed with animal matter, those species in which the latter is in considerable quantity not being numerous. The solid parts of crustacea consist of carbonate of lime, a small portion of phosphate of lime, and animal matter. The bones of fish are formed of several earthy salts and animal matter, the latter greatly predominating in those termed cartilaginous. The bones of mammiferous creatures chiefly consist of phosphate of lime, animal matter, and carbonate of lime; other substances, such as phosphate of magnesia, fluate of lime, muriate of soda, &c., being in comparatively small proportions. The enamel of teeth contains comparatively a much larger proportion of phosphate of lime and a less amount of carbonate of lime than bone. As these bodies are not equally constituted, we should not expect that if placed under equal conditions where there was a tendency of any given substance or substances to act upon them, they would retain their original component parts equally. If, for instance, all these bodies were placed in a siliceous sandstone, and water percolating through the rock were so charged with carbonic acid that it took up carbonate of lime in its passage and carried it away in solution, we might have the shells entirely dissolved, leaving only their casts, while the bones and more particularly the enamel of teeth, would suffer comparatively little destruction.

The extent to which matter in solution percolates through rocks has already been noticed, but it is nowhere made more apparent than in the various conditions of organic remains. We frequently find forms of shells which, from all analogy, we may fairly consider to have once been composed of carbonate of lime, now composed of silica and sulphuret of iron. We cannot for a moment suppose that the animals which produced these forms secreted sulphuret of iron, or nothing but silica, as molluscs now do carbonate of lime. We therefore infer that there has been a substitution of silica or sulphuret of iron for carbonate of lime, which we consider to have been the matter of the original shell. We next inquire in what situations we find a substitution of either of these substances for the original carbonate of lime; and we find that silica is most frequently substituted in siliceous rocks, and sulphuret of iron where iron pyrites are much disseminated. We thus arrive at something like cause and effect. It would follow that the greater the porosity of the rock, all other things being equal, the greater the probability of change; and, conversely, the more impervious the rock to the percolation of water, the less the probability of alteration in organic remains. We should therefore expect that organic remains in porous siliceous sandstones would be liable to greater change than in stiff clays; as we find to be the fact.

Various other circumstances must necessarily be taken into account. Time is an essential element, and one that cannot be neglected: effects that cannot be appreciable in short periods, become considerable after the lapse of ages. Hence, all other things being equal, we should anticipate that the greatest amount of change, if such be produced, would be exhibited in the oldest fossiliferous deposits. Again, the conditions under which a given deposit may exist may differ from different states of the earth's surface in the area where it occurs; it may be covered by one kind of rock at one time, and by another at another, the former having been removed by various denuding causes. Any districts where newer rocks have covered older deposits, and the whole has subsequently been so acted on that portions only of the newer remain here and there on the older rocks, will show the effects that may be thus produced. Let us, for instance, take the lias or inferior oolite of part of Dorsetshire. These rocks have subsequently to their formation been covered by large overlapping masses of green sand, chalk, plastic clay, and perhaps many other supracretaceous deposits. A large proportion of these overlapping rocks is now swept away, and the lias and inferior oolite exposed over comparatively considerable areas.

It will be evident that water percolating to the inferior oolite or the lias, under the different conditions above noticed, may be charged with different substances in solution, and the effects be modified accordingly. At the present day, the effects may be different at those places where water percolates through the superincumbent chalk and green sand still left on the tops of the high hills, from those where the rain-water falls immediately on the inferior oolite or lias. The lias might at first sight be considered as offering great resistance to the percolation of water; and so it and all other similarly constituted rocks do offer resistance; but that water permeates among them, though not so rapidly as in more porous rocks, is certain, and the amount of that substance disseminated is considerable, as may be readily ascertained by weighing a piece of lias marl when first taken from the rock, (at some depth,) and again weighing it after the moisture has been expelled by heat,—at least in a great measure expelled, for to effect this completely is difficult.

The constant solution of matter by the percolation of water through rocks is proved by the chemical composition of spring-water, evidently thus derived, which no one would consider free from foreign matter, and certainly no chemist would employ as pure water. Organic remains, like the other component parts of a rock, are exposed to this action of water, which will produce greater or less effects according to circumstances. *A priori*, we should anticipate that shells imbedded in limestone rocks would suffer the least apparent change, as they are surrounded by a substance similar to that of which they are composed; and that if the original carbonate of lime be removed by the percolation of water, the hollow or mould would be again filled with carbonate of lime, more or less crystalline according to the time required to complete the infiltration. Now this seems to be actually the case in nature.

It is generally considered that sulphuret of iron is a substance insoluble in water, and hence that the organic remains frequently composed of it in marls and clays have suffered a chemical change otherwise than by infiltration. There is much difficulty in considering this view correct, particularly when we regard the manner in which the sulphuret of iron fills up the moulds or cavities left from the removal of the shell by solution. All the appearances seem so closely analogous to those of the infiltration of silica, which no one seems to doubt, that there is considerable difficulty in believing that the cavities left by the shell have not been filled in both cases by the infiltration of water containing the matter in solution now deposited in the cavity. It by no means follows because sulphuret of iron may not appear soluble when experimented on in our laboratories, that it really is not so in such small quantities as to be appreciable only after a great lapse of time; or that conditions may not exist in natural processes favourable to such a solution which are either not attainable or have not hitherto been employed in those which are artificial.

In some fossils, portions are composed of one substance and portions of another. In ammonites, for instance, we sometimes find the walls of the cells formed of sulphuret of iron, while the cells themselves are filled with carbonate or sulphate of lime. The question then arises as to which of the substances first occupied their present respective places, and is one often very difficult to answer, because we are not certain how many changes may have been produced. It would, however, appear certain in such cases as these, where a part corresponding exactly with a previous organic form is replaced by one substance only, that there had not at one time been a complete cavity in the rock corresponding to the volume and form of the whole ammonite, but that as the cells were in all probability originally hollow, they were first filled, and that the original walls of the chambers were afterwards removed and replaced by sulphuret of iron. This view is borne out by the appearances observed when the original wall of the cells remains little changed, and the cells are only partially filled up, the cavity being lined with crystals. commonly of carbonate of lime, resting against the walls of the chamber; it appearing requisite that the chambers should be filled by some foreign matter before any substance can replace the walls exactly. We have often had occasion to observe, when these chambers were only partially filled and were broken into, the whole being in place in the rock, that they contained water; a condition highly favourable to the development of that crystallization which we often observe in them.

There is another kind of mixed matter in organic remains which shows a change in the nature of the infiltration into the same cavity, in the manner of some agates: it is one, however, by no means very common. After the carbonate of lime of the shell has been removed, and an exact mould of it left, infiltration of matter has taken place and lined, if not entirely, at least to a considerable extent, the outer sides of the cavity. A change has then been effected in the matter infiltrated, and a new substance fills the interior of the cavity. This is well illustrated by some of the gryphites found in calcareo-siliceous rocks at St. Hospice, near Nice, the outer and thinner parts of the shell being replaced by a chalcedonic substance and the interior of the thick parts by crystallized carbonate of lime *.

We generally find that the phosphate of lime contained in bones retains its place with considerable tenacity, even in rocks of very great antiquity, precisely as we should expect, since phosphate of lime is so little soluble. Consequently we rarely find bones completely replaced by silica, crystallized carbonate of lime, sulphuret of iron, or sulphate of lime, substances which so commonly replace shells. All these substances occasionally enter into the structure of the bone, replacing the animal matter, but the phosphate of lime remains. We thus sometimes see fossil bones strongly impregnated with silica and sulphuret of iron according to the different conditions under which they have been placed. The latter frequently enters into the composition of bones imbedded in clays, more particularly when sulphuret of iron is common in the rock. Saurian remains under such circumstances often contain a large quantity of iron pyrites, but the structure of the bone is not destroyed, the sulphuret of iron having only replaced the more soluble matter of the bone.

* It might be supposed there would be a difficulty in the infiltration of carbonate of lime under such circumstances, but similar facts, though rare, are observable in agates, which seem to require the same explanation.

In some rocks, cavities once occupied by shells are common, and we have only what are termed *casts* *. These cavities occur both in calcareous and siliceous rocks, and also in certain argillaceous slates, but they are rare among clays and marls of the more modern fossiliferous rocks. They may perhaps be considered most frequent among siliceous rocks. Mr. Conybeare long since noticed some singular effects, resulting from the disappearance of the matter of the shell, in the chert of the green sand near Lyme Regis †. The shells, prior to their being entombed in the rock, were pierced in various directions by parasitical creatures in the manner so frequently observable in the common oyster. After, therefore, the shells were imbedded, and when the silica arranged itself in the shape of chert beds, it infiltrated into all the little cavities formed by the parasitical creatures, so that the carbonate of lime of the shells being eventually removed, the radiated and other forms of these little passages and chambers appeared like stars and fine filaments of silica in the principal cavity once occupied by the shell. Such appearances are not uncommon, particularly in chert.

Of all the substances which have either replaced or covered up organic matter, silica affords us the most beautiful and instructive results. It has insinu-

* These cavities, which are more properly moulds, for the most part represent the external parts so accurately that very elegant casts may be obtained from them. The substances that can be employed are necessarily various, but very beautiful casts can be taken with little trouble in wax. When we consider the facility with which such casts may be obtained, it appears somewhat singular that our collections do not contain more of them, particularly as they would often be important.

† Geological Transactions, 1st series.

ated itself into all the little pores, and we thus see vegetable and sometimes animal structure in a manner scarcely attainable by any other means. Some of the fossil vegetables from Antigua are particularly remarkable in this respect. There are specimens where the silica has protected the fronds of palms, or something analogous to them, in their unexpanded state, from decomposition after it had commenced. Now as these plants from Antigua are of kinds indicative of a warm climate, this infiltration of silica arresting decomposition must have been rapid, and the plant must have passed very quickly from the conditions under which it grew, to those where it received at least its first protecting coat of silica *.

A rapid infiltration of silica would also seem necessary for the preservation of those alcyonic bodies found in the flint and chert of the cretaceous series. The structure of these bodies is ill fitted to resist pressure, and the state of their preservation is commonly such as to justify us in supposing that they have never been exposed to it, at least until mineralized in the rocks where they are now found; for their original texture is often as beautifully preserved in silica as if this substance were suddenly introduced into, and filled up a sponge under no greater pressure than that of the sea or the atmosphere.

All those who have collected organic remains from the rocks themselves must have observed how frequently they are compressed in marls, clays and

* Some beautiful specimens illustrative of this fact were three or four years since sent with numerous fossil silicified plants from Antigua, and are now in the collection of the celebrated botanist Dr. Robert Brown.

argillaceous slates, compared with the fossil contents of limestones and sandstones. This is what we should expect; for the two latter, more especially the limestones, could have received little compression from superincumbent weight after they were first deposited, compared with the mud from which the clays, marls, and fossiliferous argillaceous slates have been derived. When we recollect that every rock of the density of 2·6 and 500 feet thick presses with a weight equal to about 577 lbs. upon the square inch, we may readily conceive that the particles of mud beneath such a weight would be squeezed together; and yet it may be considered small when compared with that to which many marls and clays must have been subjected. Shells and other organic remains contained in the mud must give way, and become compressed or fractured, as the case may be, and as we now find them in such situations.

In limestones, on the contrary, organic remains generally retain their original forms, and there is little evidence of compression, except in seams of marl which may be interstratified with them. The calcareous matter appears for the most part to have been deposited among the organic exuviæ from aqueous solutions of carbonate of lime, in the manner we observe on the small scale in different situations. Solidity probably accompanied the envelopment of the organic remains, and hence they could not be compressed. The same also has probably, to a certain degree, happened with the exuviæ in sandstones, the rock containing them not being capable of compression to a great extent. Variations in this respect may often be seen well illustrated in alternations of sandstone and shale, or in the mineralogical passages of one into the other. In some cases, also, we seem to arrive at the relative data when the compression of organic remains took place in shales and marls; for calcareous and other nodules occurring in them present us with uncompressed, or but slightly flattened, organic remains, while other remains, precisely analogous and evidently entombed at the same time, are compressed in the shale itself. We have had occasion to remark that such nodules are aggregations of similar matter which have separated from the mass of mechanical rocks after deposition. Hence the compression in such cases took place after these aggregations of similar matter had been effected.

Vegetables have been entombed in various ways; some appearing to retain the places where they grew, while others have been broken into fragments, floated about and drilled by the *Teredo*, or analogous creatures, before they were imbedded in rocks. Others again are accumulated in vast abundance in particular areas, producing that highly valuable mineral, coal. Fossil vegetables possess great interest from the conclusions that may be derived from them. Those which retain the positions in which they grew afford evidence of their tranquil envelopment by the matter which now surrounds them, while those which occur in detached fragments show that they have been transported. We necessarily infer the transporting power to have been moving water, and we judge of the amount of transport by the condition of the fragment. Thus when we see a multitude of delicately preserved leaves, we infer a small amount of transport, and their quiet envelopment before decomposition took place. Changes in the component parts of vegetables have been effected according to circumstances, but the carbon has most frequently remained firm, so that fossil vegetables are generally carbonaceous. Infiltrations of silica, carbonate of lime, and other substances, into the pores of the plants, have sometimes taken place to a great extent, and occasionally there appears a complete substitution of new matter for that of which the vegetable was originally composed. When, however, we view the mass of fossil vegetables generally, a large proportion retain their carbon.

The present condition of organic remains must depend upon the circumstances to which they have been exposed, both previous to and after their deposit in the various rocks where they are now detected. Without them we should have never known that any other than the present animals and vegetables had existed on the surface of the earth. By them we possess the most decided evidence that existing animals and vegetables have been preceded by others which are not now discovered on our planet. We might, indeed, from the superposition of different mineral masses, learn that detrital and other matter had been accumulated at different periods on given parts of the earth's surface, and we might infer from the structure of such masses that they required a greater or less amount of time for their production; but a knowledge of the possible variation of climate in any given area, and of the existence of animal and vegetable life previous to and different from the present, could never have been acquired by such means. It is to organic remains we turn for such information; they teach us that man is a comparatively recent creature on the face of the globe; that creation has succeeded creation on its surface; that life has existed on it from remote geological epochs; that climates have varied over the same areas; and that there has been no stability in the modifications of animal and vegetable life since it was first called into existence on the surface of this planet.

CHAPTER XIV.

THE rocks of which the mineral crust of the globe is composed are necessarily divided into those of aqueous and igneous origin, from the two agents known to us as capable of their production. The terms 'stratified' and 'unstratified' have been commonly considered as respectively synonymous with 'aqueous' and 'igneous'. Practically, this division is highly valuable; but theoretically, it is not so satisfactory, at least, if we are to infer that all rocks divided into tabular masses, one resting on another, must have been deposited either chemically or mechanically from water. We have seen in the case of the Cornish and Dartmoor granite (p. 104.), that there are cleavage planes, dividing the masses into tabular portions and which might pass for strata, if there were not other systems of cleavage planes, dividing the whole into prisms. Scarcely any geologist will at the present day suppose that these granitic masses have been deposited from water; and if there should be any who do so, they certainly will not imagine that the deposit has been effected prism by prism, as men erect a building. Yet each of these planes tends to divide the granite into tabular masses, the most constant fissures being perpendicular with a direction from N.N.W. to S.S.E.; and if the others were not apparent, the whole might be termed stratified. Few geologists would now apply this term to the masses in question; but there may be equivocal cases in other rocks and situations, where cleavage might not readily present itself as a sufficient explanation.

Basalt may be divided into beds, and yet be of igneous origin. This is shown on the coast of Ireland near the Giant's Causeway. That this mass of basalt is igneous there is abundant evidence, that exhibited at Kenbaan, where disrupted masses of chalk have been caught up in the fused rock, being particularly striking. Now the whole range of beautiful cliffs extending from the Causeway to Dunseverie Castle exposes a series of beds, some of which are columnar, others amorphous. They form a great curve, rising above the level of the sea at one place, and descending into it in the other, the Giant's Causeway itself being composed of a columnar bed where it cuts the sea-level. It would appear that some of the basalt had become columnar from causes which did not always obtain, so that the whole was not thus characterized. As the lines dividing the columnar from the other portions are so clearly marked, we may infer that the whole mass of basalt, at least in the particular part of which this portion of the coast is a section, was not in fusion at the same time, but that there were surfaces produced by successive layers of rock formed at different intervals. Without this hypothesis, there seems a difficulty in explaining the columnar character of some layers, and not of others, and the general appearance of the whole. It does not, however, in the least follow that the whole is not of igneous origin, but merely that parts of the great basaltic tract of the North of Ireland may have been produced after others, precisely as one lava current may succeed another in a volcano, and yet the whole be of one geological epoch. The only difference being the tabular character of the Irish basalt and the linear appearance of lava thrown out from a sub-aërial volcano *.

The appearances exhibited at Staffa are precisely analogous. The trappean rocks of this island, so celebrated for its beautiful cavern, form three distinct beds, the lowest being a kind of trappean conglomerate, the central being columnar, and the upper amorphous and irregularly columnar. These beds, as no doubt is very commonly the case with those of similar rocks, are of irregular thickness; but they dip as a whole, according to Dr. Macculloch, at an angle of 9° †. It is this dip which brings the principal bed of columns within the action of the breakers, enabling them to hollow out and form caverns where there is less resistance to their action than in other places. The columns being jointed at Fingal's Cave explains, as Dr. Macculloch has observed, the formation of that fine cavern ‡. Igneous rocks may therefore be stratified, that is, rest in tabular masses upon each other, such tabular masses being in all probability produced at different intervals of time.

* When considerable masses of this basalt have been under equal circumstances at the same time, we find equal effects. Thus the columns of basalt at Fairhead must have been produced while the whole mass of basalt at that place was under equal conditions. They are the largest which have hitherto been noticed in any part of the world. The recent and accurate measurement of them by Lieut. Larcom, R.E., of the Ordnance Trigonometrical Survey of Ireland, gives 317 feet for the height of the principal columns of Fairhead. The sides of these enormous prisms sometimes measure five feet, according to the same gentleman.

† Western Islands of Scotland. ‡ *Ibid.*

As, therefore, sheets of igneous rocks, of greater or less area according to circumstances, may cover pre-existing sheets of similar rocks, and the result be stratification, that is, the successive deposition of one tabular mass of rock upon another at distinct intervals of time, we should be careful not too far to couple stratification with aqueous deposition. No doubt in the great majority of cases the one has resulted from the other, but we must not hence infer that the former can arise *only* from the latter. We have elsewhere* retained the division of rocks into Stratified and Unstratified, because it is exceedingly useful for ordinary purposes, and because it seemed convenient to do so until better terms be supplied. To have divided rocks into Igneous and Aqueous, would have been to prejudge every question connected with their origin; and although we should probably not err greatly in referring numerous rocks to the one or the other, there are many the origin of which we must prejudge, if we attempt, in the present state of geology, to arrange the mineral crust of the globe under such divisions. For the same reasons which induced us formerly to retain the terms 'stratified' and 'unstratified', we do so now; wishing it to be understood solely as a division of convenience for the purpose of considering rocks in their order of superposition and relative age. With this understanding, therefore, we proceed to notice those which we have elsewhere classed under the head of 'Inferior stratified or non-fossiliferous rocks'.

These constitute the lowest of all stratified rocks,

* Geological Manual, p. 37.

and in them no organic remain has ever yet been detected. They are crystalline, pass much into one another, particularly in the lines of strike or direction of the beds, occur in no determinate order, and in all probability constitute a large volume of mineral matter, supporting the great mass of stratified rocks which appear on the surface of the globe. Their position relatively to the other stratified rocks, and their crystalline character, naturally suggest the idea that the one is somewhat dependent on the other. Wherever found, and they have been detected and examined in various parts of the earth's surface, their general character is the same. We hence infer a considerable uniformity in whatever general cause may have produced them at the geological epoch when they were formed. We thus have an inferior position, crystalline arrangement, and a general uniform structure, as characteristics of these rocks. We may therefore infer that there was a very general common cause for their production, and this necessarily leads us to a generally uniform condition of the earth's surface at this early period.

It has been supposed that gneiss, mica-slate, hornblende rocks, chlorite slates, and others of the inferior stratified rocks were deposited in the same manner as the beds of sands, mud, pebbles and limestones, now forming, and that their present appearance is due to the action of great and long-continued heat beneath, which caused the crystalline arrangement of particles we now see, the organic remains, where they existed, having been obliterated by these means. This hypothesis necessarily requires preexisting solid matter from which to derive the detritus and upon which to deposit it; so that we do not see what is gained by it in support of the opinion that such causes only as we daily witness, or rather their effects, can explain all geological phænomena. But waiving this consideration, it is not apparent how the effect required is to be produced in the manner supposed. We have no reason to imagine that great heat prevails at the bottom of the actual ocean; if such were the case, it would be made known by the temperature of the surface water. Now certainly no experiments made on the temperature of the ocean at different depths would lead to such a conclusion, while they might do so to one quite the reverse. Hence, supposing detritus carried to the lowest depths of the ocean, where in all probability it can never get from the want of a transporting power under ordinary circumstances, it does not appear how it is to be heated there, except by heating and perhaps fusing the preexisting bottom. Granting, however, that by some means or other great heat is applied to these detrital rocks,—and under the theory of Mr. Babbage, if the bottom of a sea be covered by detritus, and central heat exist, there would be a tendency of heat to move upwards in the mass, so that the more the detritus was accumulated, the more the general heat would move upwards, and therefore detrital rocks deposited on a comparatively cold bottom might eventually be exposed to a high temperature, and consequently be altered,—we are still at a loss to understand how such exceedingly bad conductors of heat could become altered to the great depth which would be requisite.

That mechanical rocks have been altered in the

vicinity of igneous rocks, argillaceous schists, and others assuming the appearance of mica-slate, gneiss, &c., is well known, and this has probably given rise to the whole hypothesis; but there must necessarily be a limit to such alterations in proportion to the power of the stratified rock to conduct heat, and to the volume and temperature of the heated mass, brought into contact with it. There is, however, a vast difference between a power of this kind and one capable of changing any amount of deposits, such as are now daily forming, into all that immense mass of inferior stratified rocks known to exist in various parts of the earth's surface. We should conceive that, with such bad conductors as rocks, the distance cannot be very considerable from the igneous heated matter at which detrital strata could be altered without fusion and consequent total loss of stratification. To alter, therefore, such a mass of matter as that composing the inferior stratified rocks, seems scarcely possible in the manner supposed. Let us, however, consider, for the sake of the argument, that these deposits could permit the heat to pass through them in such a manner that the particles of matter arranged themselves in a crystalline form, and that stratification was preserved. We should expect to find that the original detrital deposit, and the rocks produced by the crystalline arrangement of the same particles, were nearly, if not entirely, composed of the same elementary chemical substances, except perhaps near the contact with the heated matter causing the alteration.

The great distinctions which immediately strike us in the two kinds of rocks, are the large comparative proportions of carbon and lime in modern deposits, particularly if fossiliferous, and the great scarcity of the former and the comparative rarity of the latter in the rocks under consideration. Shells and other calcareous organic remains, should they alter their form, would become crystals of carbonate of lime, and other obvious changes would be produced with the bones of saurians, and other remains, for we are not to suppose fusion but simple alteration in the position of the component particles, the carbonic acid retaining its place. Independently, therefore, of various difficulties, and others might readily be mentioned, the products required would differ chemically from those which would be obtained under this hypothesis; and the same would be the case whether we substituted the supracretaceous, the cretaceous, the oolitic, the red sandstone, or the carboniferous groups, instead of the rocks now forming. More might, perhaps, be said for the mass of the grauwacke group, but we should still have serious difficulties. From what is here stated, there is little probability of the inferior stratified or non-fossiliferous rocks having been deposits similar to those now forming, altered by heat; and when we study the relations of the former to each other, the probability appears, if possible, still less.

When the inferior stratified rocks are developed on the large scale, we generally find that gneiss and mica-slate are the most abundant rocks, constituting the far larger proportion of the mass. Hornblende rocks seem to succeed in general importance, then the eurites and quartz rocks, talcose slates, chlorite slates, and argillaceous slates. Limestones and dolomites are so sparingly disseminated amid the mass of the rocks under consideration, that their relative volume must be exceedingly small. They are, however, important in a theoretical point of view, as they show that the inferior stratified rocks were not altogether composed of various silicates, and that carbonates were formed among them, though in exceedingly small comparative quantities.

We have elsewhere* calculated the proportions of elementary substances entering into the general composition of the rocks known under the names of gneiss, mica slates, &c. Such calculations are no doubt mere approximations, but they are sufficient to show us the leading chemical distinctions between the rocks in question, both as regards themselves and those which have been since formed. Let us suppose that a large district is formed of gneiss and mica slate in equal proportions, $\frac{2}{6}$ths of the gneiss being composed of equal parts of quartz, felspar and mica, $\frac{1}{6}$th of the same rock of equal parts of albite, quartz and mica, and $\frac{3}{6}$ths of mica slate composed of equal parts of quartz and mica, the latter mineral being of the fluoric acid kind, we should have, as the mean of one hundred parts of the whole volume of such inferior stratified rocks developed in the district,

Silica	71·87	Lime	0·25
Alumina	14·09	Oxide of iron	3·45
Magnesia	2·08	Oxide of manganese	0·30
Potash	5·73	Fluoric acid	0·45
Soda	0·55	Water	0·86

It will be obvious, that the relative proportions of the different rocks of the inferior stratified and non-fossiliferous group being fairly estimated in a given

* Geological Manual, pp. 440—443.

district, an approximation towards the elementary substances contained in the mass may, in this way, be calculated with sufficient precision for various important inferences.

Taking gneiss, formed of equal parts of quartz, felspar and mica, to be composed, as elsewhere calculated*, of silica 70·06, alumina 15·03, magnesia 1·66, lime 0·37, potash 7·92, oxide of iron 2·97, oxide of manganese 0·20, fluoric acid 0·36, and water 0·66, and making this rock a standard of comparison, we obtain the following approximative differences as to chemical composition between it and the rocks enumerated beneath.

	A.	B.	C.	D.	E.	F.	G.	H.	I.
Silica	+1·86	+3·01	− 8·12	−15·20	− 6·35	+ 8·38	+5·18	+5·14	+11·96
Alumina	+0·17	−1·95	+ 0·42	+ 0·53	− 6·08	− 9·63	−8·44	−0·03	− 5·56
Potash	−3·55	−1·86	− 4·55	− 1·09	− 7·14	− 7·92	−3·37	−4·52	− 1·09
Soda	+3·31								
Magnesia	+0·04	+0·83	0·00	+ 7·73	+ 5·62	+11·54	+7·60	+0·74	− 1·66
Lime	−0·12	−0·20	+ 0·08	+ 6·92	− 0·12	+ 1·63	−0·04	+0·83	+ 0·01
Fluoric acid ...	0·00	+0·18	0·00	+ 0·39	− 0·36	− 0·36	−0·36	−0·36	− 0·36
Ox. iron	−0·21	+1·11	+11·75	+ 1·06	+12·34	+ 1·08	−1·89	−2·97	− 2·60
Ox. manganese	+0·06	+0·10	+ 1·03	− 0·09	− 0·20	− 0·20	−0·20	−0·20	− 0·20
Water...........	−0·21	+0·34	0·00	− 0·66	− 0·66	+ 0·84	+1·34	+0·84	− 0·66

A. Gneiss, composed of equal parts of albite, quartz and mica; B. Mica slate, of equal parts of quartz and mica; C. Mica slate, of equal parts of quartz, mica and garnet; D. Hornblende rock, of equal parts of hornblende and felspar; E. Chlorite slate, of equal parts of chlorite and quartz; F. Talcose slate, of equal parts of quartz and talc; G. Protogine, of equal parts of quartz, felspar and steatite; H. Eurite, such as that of Nantes; and I. Quartz rock, composed of equal parts of quartz and felspar.

* Geological Manual, p. 440.

We thus see that the modifications of the principal rocks of this class depend upon a few different combinations of a small number of silicates. These modifications are extremely interesting to trace in nature, where we often find the inferior stratified rocks pass from one into the other in the most gradual manner, more particularly in the line of their strike or direction, so that it is scarcely possible to see where the change precisely commences or terminates. The gradual increase or decrease of some one or other of the component silicates is quite sufficient to alter the appearance of the rock, converting it from one known compound into another. These passages are, however, often so singular, that to apply to them any of the names usually assigned to given mixtures of these various silicates seems impossible, however convenient it may be to affix given names to the more marked compounds.

Notwithstanding these minor changes, the common character of the whole mass is so striking that we are naturally induced to search for some simple cause productive of its common origin. If the theory of central heat be founded on great probability, as it seems to be, we should look for this cause in some one condition of the earth's surface at a given epoch which should present us with the uniform conditions required. Now, under this theory, there must have been a time when the mineral crust first became solid, and there must also have been a time when the mineral surface, being sufficiently cool, permitted the existence of water in its liquid state. These two events could scarcely have been contemporaneous, because we should infer that the superficies of our spheroid, when first solid, would still possess too high a temperature to permit water to rest in its liquid state upon it. A time, however, must come when the heated rock would merely allow the water to remain liquid upon it, and there would be a condition of things in which a great mass of heated water existed on the face of the globe, which, driven off by evaporation, condensed in the atmosphere, and fell again into the mass beneath. Such a condition of things would necessarily be one in which neither animal nor vegetable life, analogous to that now found, could exist. Hence no organic exuviæ could become entombed in any rocks which could result from this state of the earth's surface. It would also be one highly unfavourable for the production of carbonate of lime, since the carbonic acid would be driven out of the water, and, consequently, no carbonate of lime could exist in solution. We should moreover recollect that lime itself is less soluble in hot than in cold water, for Dalton has found that one grain of lime requires for solution

778 grains of water at 60° Fahr.
1270 grains of water at 212° Fahr.*

Lime itself would, therefore, be sparingly dissolved, and we should scarcely expect to have any carbonate of lime in solution, and consequently none could be thrown down as a deposit. The case would, however, be very different with the silicates; for numerous thermal springs afford evidence that hot water is favourable to the solution of silica.

We have next to inquire how far detrital rocks

* Turner's Elements of Chemistry, 4th edition, p. 467.

might be produced under this condition of the earth's surface. Surface currents are produced by winds, the friction of which on the water causes a forward motion in their direction; the uniform temperature of an atmosphere charged with hot aqueous vapour would not appear favourable to the production of winds causing currents of geological importance. The action of the tides, however, due to other causes, would be in full force; so that if there were inequalities in the earth's solid surface producing shallows and even land above the water, as when we compare the crust of the globe with its volume we should expect must have been the case, there would be tidal currents capable of transporting substances to distances proportionate to obvious circumstances. If parts of the solid surface rose above the ocean level, the aqueous vapours would probably condense upon them, and running waters would be the result. We might, therefore, anticipate that mechanical rocks could be produced under these circumstances, and therefore should there be traces of mechanical rocks among the inferior stratified class, there would be nothing very remarkable in the circumstance under this hypothesis. We may, indeed, go further, and consider that such rocks would very probably exist; for we may infer that, in such a condition of the earth's surface, disruptions of the thin solid crust would be frequent, and would be productive of great waves and currents of the water beneath, and in which the fractures and consequent agitation were effected. Such currents would necessarily tend to abrade and transport solid matter.

Many of the inferior stratified rocks have evidently

been chemical products, and some perfectly resemble granites and greenstones divided into beds. This is particularly remarkable in that kind of gneiss which occurs in thick beds, and contains large disseminated crystals of felspar or albite, the plates of mica in which are not arranged parallel to the plane of the strata. Some beds of hornblende rocks differ little from beds of greenstone. We have had opportunities of seeing how both these kinds of rock finally pass, in their lines of strike or direction, into common gneiss or hornblende slate with mica. The former by the gradual arrangement of the plates of mica into one plane, and the latter by the acquisition of mica, which, when sufficiently abundant, converted it into a highly schistose rock. These are very marked differences from the facts observable in common granite or greenstone, and point to at least a considerable modification of the circumstances under which they have been produced.

We thus obtain an absence of organic remains, the probable rarity of carbonate of lime, the prevalence of silicates, and the possible mixture of chemical and mechanical deposits. We should expect the inferior rocks also to correspond in their general chemical characters with those igneous products which have been thrown out at different geological epochs on the surface of the globe. Now this resemblance does exist, more particularly among the older igneous products, precisely where we should expect to find it. All the conditions supposed would be highly favourable to a mixture of the igneous rocks of the period with those thrown down from the heated ocean. And we should recollect that this ocean would contain numerous substances in solution, part of which remain until this day. It is needless to point out the coincidence of these results with the facts observable among the inferior stratified rocks. If our planet were once fluid, as there is every reason to suppose it must have been, and if that fluidity were igneous, as in all probability it was, we can scarcely consider but that a condition of things somewhat similar to that noticed above should have obtained; and, therefore, we should expect to find a series of rocks formed at that period differing from those produced at later epochs, the most marked character of which should be the absence of organic remains.

CHAPTER XV.

We now proceed to consider briefly some of the circumstances connected with that important division of rocks which we have termed fossiliferous, and which show, by the organic remains entombed in them, that life was created on the surface of our planet before they were formed. It has been found that the European fossiliferous rocks, those which have hitherto been best examined, may be conveniently divided into groups, as they present differences, more particularly in their organic contents, which would appear to authorize such subdivisions. As the relative order of the series is never inverted, though parts of it may be, and are often absent, we have to a certain degree a record of geological events, as far as regards the appearance or disappearance of particular animals and plants over given areas. When known deposits are absent from the series, it becomes highly important to discover, if possible, whether they ever have or have not existed in the particular area examined. A given rock B may have been deposited upon another A, and have subsequently suffered removal by denuding causes, at least in the area examined, so that another rock C may be deposited directly upon it. When this has happened, we should expect to see marks of erosion or some other evidence of the action of denuding causes on the upper surface of A; and this evidence we frequently obtain. There would necessarily be in such cases an uncertainty as to the succession of animals or plants which may have occurred in such part of the earth's surface. Now it may, and often has, happened that a rock such as C has been deposited on a part of the surface of A which has never been covered by B. So that should we be desirous of knowing in what manner given contemporaneous and fossiliferous rocks may have been distributed over a particular part of the earth's surface, and consequently the kind of animals and vegetables which succeeded each other, we must carefully examine the conditions under which two fossiliferous rocks may rest upon each other. This is no doubt often a work of considerable labour, but much depends upon it. For instance, the lias rests quietly in many parts of England upon the red sandstone group, in the upper part of which group in the same country there are few or no traces of organic remains, and no decided development of limestone. It hence might be, and was long, inferred that there was no deposit containing organic remains between the lias and the magnesian limestone, and, consequently, that the kind of animals and vegetables whose remains were discovered in the lias succeeded those whose exuviæ were entombed in the magnesian limestone or zechstein. It subsequently appeared that this had not been the case; for a remarkable calcareous rock, the muschelkalk, found in the upper part of the red sandstone group in parts of Western Europe, extending from Poland to the Mediterranean coast of France, contains the remains of creatures differing generally from those both in the lias and in the magnesian limestone. Hence the different animals and plants, known to have existed from their remains found in the two latter rocks, did not immediately succeed each other in the European area. Others of a different kind flourished at an intermediate period, and either did not live over that part of the area, where England now rises above the surface of the sea, or there was something in the circumstances connected with the deposit itself, in the same situation, which did not permit the preservation of any part of their remains.

Any estimate, therefore, of the succession of animals and plants which have existed on that part of the earth's surface now known as Europe, requires much care and great caution, more particularly when we attempt, from the remains entombed in rocks, to sketch their distribution at a given geological epoch. If this requires great care, is not still greater caution required when we draw inferences respecting the general condition of life over the whole surface of the globe at any one geological period?

A large portion of the fossiliferous rocks have been produced from the deposition of matter mechanically suspended in water, and, generally speaking, the finer the deposit, the greater the appearance of the tranquil envelopment of the organic remains found in it. Subsequent causes may have compressed and fractured the remains in place, but all other things being equal, a fine deposit, such as we may suppose to have once been mud or silt, but which is now marl or clay, is that which among the mechanical rocks affords the most perfect remains. As we have previously remarked, compact limestones may be considered more as chemical than mechanical deposits, though strictly speaking the carbonate of lime thrown down as a precipitate from water is insoluble in the latter at the time, and is therefore mechanically suspended in it; so that we should expect organic remains to be well preserved, as they are, in such limestones, the fine particles of carbonate of lime quietly entombing them.

A considerable thickness of sandstone rocks, the grain of which is often exceedingly fine, and of associated argillaceous slates and limestones, constitutes the lowest group of rocks containing organic remains with which we are acquainted. This is the grauwacke group. There is abundant evidence that moving water transported detritus and deposited it in beds; these latter would, however, if measured perpendicularly, present us with an aggregate thickness, so considerable, that we are led to inquire if there may not be something deceptive in the appearances observed. We have seen (p. 51.) that beds of sand may be formed above each other at angles up to 20° or 30°, and cover a bottom of moderate depth in such a manner, that if they were measured as we do the grauwacke beds, the result would be extremely erroneous. Now as the grauwacke measured in this manner usually gives us many miles of thickness in the different districts where it is observable, it would require either a depth of ocean we should have some difficulty in conceiving ever to have existed on the earth's surface, or a gradual descent of bottom, as the detrital matter was accumulated above, to an equal amount, in order to account for the thickness thus estimated, on the supposition that the beds have been successively deposited in a horizontal manner. When, therefore, we have an explanation which does not require more than the conditions now existing off the edges of soundings round various coasts, it seems fair to infer that much of this apparently great thickness may be due to causes so far similar, that the detrital matter has been gradually moved forward on the bottom, and thus thrown over into deep water, where, particle supporting particle, the resulting beds were inclined at angles which, under favourable circumstances, approached 20° or 30°.

Among these beds we begin to find a greater abundance of carbonate of lime, which, though still in small quantities compared with the amount contained in the rocks somewhat higher in the series, shows a certain change in the conditions under which rocks were then produced. Now though it is not always the case, yet we generally find organic remains in the calcareous beds, and viewing this group as a whole, it is principally in the grauwacke limestones that the greater quantity of organic remains is detected.

The difficulty of separating the grauwacke group from the inferior stratified rocks is extreme, for where good sections have presented themselves, and we have the lower and not the upper part of the grauwacke series, the two classes of fossiliferous and nonfossiliferous rocks appear to graduate into each other, principally by alternations. This is precisely what we should expect if the theory of central heat were correct; for the waters of the globe would, as the whole surface became cooler, rest more readily on the surface of land, and there would be a greater tendency to produce detritus, rivers and other agents of that kind being more abundant. Solutions, also, which depended on the heated state of the waters would become less frequent as these cooled.

As from all analogy we cannot conceive animal life

to exist in greatly heated water,—which, moreover, would in proportion to the heat in it tend to drive atmospheric air out of it,—we must assume a moderately cool ocean as a necessary condition for the existence of marine creatures. When it became moderately cool above, there would be decreasing temperatures beneath, so that there would be different levels, which might be tenanted by differently constituted creatures.

When we compare lists of the organic remains entombed in the grauwacke group with existing marine life, as far as it is known to us, we find many genera common to both. Now if we infer that these genera have in general had similar habitats at the different epochs, which it must be confessed is an inference not altogether safe, we should conclude that they were either littoral, inhabitants of moderate depths, or free swimmers in the ocean. The same circumstances, connected with the medium in which they existed, must have affected the marine creatures of the grauwacke epoch as affect those of the present day. Creatures with eyes would require light; those living habitually in small depths would not support life equally well beneath great pressure at considerable depths, and different kinds of bottom would suit different creatures. The whole of the marine exuviæ observable in the grauwacke is such that we may infer it could exist at moderate depths, and that two or three hundred fathoms of water would suffice. It may be said that we cannot infer the habitat of the *Trilobites*, since nothing precisely analogous to them has yet been detected in our present seas. This may be true enough as to the kind of bottom which they may have preferred; but as they possess eyes, we may fairly infer that light was necessary to them, and that therefore they did not live at very considerable depths. They are sometimes entombed by myriads in the beds of a small district; and as they are then often complete, and have apparently been imbedded either living or before decomposition took place, we may consider that some species at least were gregarious.

It would not suit the plan of this work to enter into details respecting the organic contents of the grauwacke; but if the reader will compare the lists of organic remains stated to be contained in this rock with the table of the known habitats, as far as regards depth of water and bottom, at the end of the volume, he will probably come to the conclusion that the marine life of this early period occupied bays, creeks, shallow open water, moderate depths, and the open ocean, precisely as marine creatures do at the present day; that some genera have disappeared from the surface of the globe, while others still exist, and that many now known have not been discovered in this group.

As carbon is essential to the animal and vegetable life now existing on the surface of the earth, it seems fair to infer that it was always necessary to animal and vegetable life. Now it is worthy of remark, that with the early appearance of life the calcareous matter in rocks increased. Carbon being so exceedingly rare in the inferior stratified rocks, we are led to inquire whence it was derived. At the present day a large amount of carbon, combined with oxygen and forming carbonic acid, is thrown daily into the atmosphere through cracks or volcanic vents in the earth's crust. If there were a condition of the earth, as we have supposed there was, when the waters were so hot that they could not absorb carbonic acid, and there was neither animal nor vegetable life to appropriate a part of its carbon, any carbonic acid thrown from the interior of the earth upon its surface would remain in the atmosphere, except such portions as could form compounds with the mineral matter rising above the level of the waters and sufficiently cool for the purpose. As the general surface of the world became cooler, these conditions would necessarily change, and carbonic acid would be absorbed by waters; these would then act chemically on various substances, and among other things they could take up carbonate of lime in solution, which without this addition of carbonic acid would have been insoluble in them. A large proportion of carbon would be appropriated by the mass of animal and vegetable life when first created, and an equal volume of oxygen would be liberated for the support of the creatures then called into existence. The atmosphere would not only be thus purified by the abstraction of a certain amount of carbonic acid, but it would also be rendered more fitted for the support of life by the additional proportion of oxygen thrown into it.

Under the conditions we have supposed, we should have a sea of an uniform temperature at different levels, land rising above it in various places, and necessarily waters of different depths. Such conditions would be highly favourable to a certain general uniformity in the distribution of animal and vegetable life. It has been considered that this general uniformity does prevail in the organic contents of the grauwacke. This is certainly in a great measure true with respect to the grauwacke known to us in Europe and North America, the organic contents of which are remarkably similar; but from this circumstance it would be evidently unsafe to conclude more than that, in certain parts of the northern hemisphere, conditions for the existence of certain creatures were equal at the same geological epoch. What the conditions may have been in the tropics, and in the southern hemisphere at the same time, can only be conjecture, because the requisite data fail us. Until it was well understood that rocks of different ages may have the same mineralogical structure, grauwacke was stated to be found in many parts both of the tropics and southern hemisphere; but as yet we have no proof that the rocks so called are really such. In all probability there must have been equivalent rocks in these portions of the earth's surface, and it will be highly interesting, as their organic contents become known, to see how far they correspond with, or differ from, those of the grauwacke of Europe and North America. As we have had occasion to remark, neither general uniformity of organic contents nor mineralogical structure proves continuity of deposit; therefore we might at all times have detached masses of land around which there might be deposits perfectly distinct as masses of matter, though they zoologically resemble each other.

The recent researches of Mr. Murchison on the upper division of the grauwacke of Wales, and on the continuation of this division into the neighbouring English counties, have shown, at least as far as regards this area, that there is even a marked distinction between the organic contents of the various subdivisions of this portion of the grauwacke, so that we might expect still greater between it and the lower division. Such distinctions would suppose change of circumstances which should affect the animal life previously existing in this particular area, either causing a destruction of part of it, or the removal of such part elsewhere in search of the conditions suited to it. How far a change in the organic contents of the European grauwacke should correspond generally in this respect would depend on the similarity of the conditions to which the whole of it has been exposed; so that, *à priori*, it would be hazardous to assume that the upper grauwacke of Podolia, of the neighbourhood of St. Petersburg, of Sweden and of Norway, remarkable for the general horizontal character of its beds, presented exactly the same organic remains.

The carboniferous group succeeds the grauwacke in the ascending series of Europe*, and is so called because the great mass of European coal is included among the rocks of which it is composed. It is stated that a large proportion of the American coal is associated with the preceding group; a statement which, at all events, can have nothing improbable in it, since, if terrestrial vegetation existed during the time that the grauwacke was forming, there appears no reason why the remains of such vegetation should not be found in that group. Indeed the upper part of the grauwacke group of Europe contains beds of anthra-

* For the order of succession of the various groups of fossiliferous rocks, with descriptions of them, and of the countries where they have been discovered, the reader is referred to the Geological Manual, and other geological works.

cite with vegetable remains. M. Elie de Beaumont notices the occurrence of anthracitic coal in the grauwacke of the Bocage (Calvados), and of the interior of Brittany, worked for profitable purposes, and containing the remains of plants which do not much differ from those of the coal-measures, properly so called. The same author also refers rocks containing anthracite and fossil vegetables, forming a part of the south-east angle of the Vosges, to the same epoch. M. Virlet considers the coal of St. Georges Châtellaison referrible to the grauwacke; and Mr. Weaver refers all the coal in the province of Munster, with the exception of the county of Clare, to the same series. It is an extremely interesting point to see how far facts justify us in concluding that vegetation and animals have appeared contemporaneously, or nearly so, on the surface of the globe; and further to study, if analogy can help us, how far both may have been suited to a state of the atmosphere differing from the present in the relative portion of carbonic acid contained in it.

The researches of M. Adolphe Brongniart on fossil plants led him to conclude, that during the early period when the plants entombed in the coal-measures flourished, the atmosphere was more charged with carbonic acid than at present, aiding the development of the gigantic species, the remains of which are there detected, and also protecting them when dead from being so readily decomposed by the atmosphere. Now this conclusion is remarkable, because we have arrived at the same by a totally different train of reasoning. It has also been further inferred that the carbon abstracted from the atmosphere, gradually fitted it for the respiration of reptiles, and, finally, for mammiferous animals. If the reasoning we have employed to show the changes that should take place, under the theory of a gradual reduction of the earth's temperature, approximate towards the truth, the atmosphere would also be purified when the waters became sufficiently cool to permit the absorption of carbonic acid, and the consequent formation of calcareous deposits. We have shown that the volume of this gas locked up in limestones is enormous, being at the rate of 16,000 cubic feet for every cubic yard of limestone (p. 12.). Even existing shells must contain a considerable volume of it, for in the two valves of the great Tridacna, now at Paris, and weighing 500 lbs., there is, assuming them to consist of carbonate of lime, somewhat more than 3,250,000 cubic inches of carbonic acid, and, allowing for animal matter, about 3,000,000 cubic inches. We should thus obtain, by a given reduction of the earth's temperature, a state of things fitted for animal and vegetable life, which was not so fitted at a prior period. So long as carbonate of lime could not be freely formed in waters, the numerous marine creatures which require it in such comparatively large proportions, such as the shell-bearing molluscs, the encrinites and corals, the remains of which are found in the lowest fossiliferous deposits, could not procure it in the requisite proportions; and hence, independently of the elevated temperature of the water, and of the difficulty of procuring disseminated oxygen, circumstances would have been ill suited to their existence*.

* Though once a disputed point, it seems now certain, more particularly from the researches of Mr. Gray (Phil. Trans. 1833),

Those botanists who have paid attention to the subject of fossil plants agree in considering the remains of vegetables detected in the coal-measures as analogous to that of the tropics, more particularly of islands in the tropics. Notwithstanding differences of opinion respecting the particular existing genera to which these plants may bear the greatest resemblance, the high temperature of the climate at the time appears to be generally conceded. It has even been considered, from the great development observed to have taken place in certain kinds of plants, that the climate might have been ultra-tropical. Now it should be recollected that these conclusions have been deduced from facts observable between the parallels of 40° and 60° of north latitude; so that should there have been anything like an ultra-tropical climate in such latitudes, we should expect that a warm state of the atmosphere would be very general, bringing with it the consequences of such a state of things.

It has been observed that the grauwacke passes into the old red sandstone (as it is termed,) above it in one part of Britain, while conglomerates considered

that molluscs have the power of dissolving their shells. That several have the power of piercing calcareous rocks has been long known. As water containing carbonic acid can take up carbonate of lime which would otherwise be insoluble in it, it would appear to follow, that if the molluscs could impregnate water with carbonic acid, and apply the water thus charged to their shells or calcareous rocks, the carbonate of lime would be gradually dissolved. Now molluscs appear to form carbonic acid during respiration in the manner of creatures which breathe air; and, therefore, if they could impregnate a portion of water with this acid, which might be accomplished by confining the expired water to particular places, and, perhaps, respiring it several times, they would have the means of dissolving the calcareous matter either of their shells or of rocks.

to represent the latter rest on the disrupted edges of the former in another part of the same country; showing, in a comparatively small area, that while tranquillity prevailed and continued in one part of it, disrupting forces were in action in another, preventing the continuous and quiet deposition of detrital and other matter. Professor Sedgwick has ably pointed out the range of the old red sandstone through Great Britain; and, alluding to the absence of that deposit between the carboniferous limestones and grauwacke of North Wales, considers that the "old rocks of North Wales underwent a great movement, anterior to the period of the old red sandstone, and that by this movement the bottom of the neighbouring seas was raised out of those causes which produced the old red sandstones*." The tilted character of the grauwacke and the conglomerates here and there interposed between it and the carboniferous limestone, which are often, as Professor Sedgwick observes, of great thickness, show the application of force and the destructive action of water.

It can only be under favourable circumstances, such as those which exist in the British Islands, that the connexion between the grauwacke and the carboniferous groups can be satisfactorily traced to considerable distances. In those situations, such as the southern parts of England and Wales, probably also on the Rhine, where the two pass into each other, separation is purely artificial; for, in fact, they then merely constitute the upper and under parts of a mass of detrital matter deposited tranquilly above each

* Geological Manual, p. 391.

other, and are therefore no more geologically separable than the component strata of any given rock, such as the lias, which may differ quite as much in their organic contents. Such separation, however artificial it may be, has nevertheless its importance, as it shows the points where tranquillity has prevailed at equal geological epochs; and hence, when facts shall have been sufficiently accumulated, we may the better judge of the relative amount of areas undisturbed at the same period.

It was at one time supposed that organic remains were not found in the old red sandstone; the recent researches of Mr. Murchison in the districts above noticed have, however, shown that this rock does contain organic exuviæ, though they may, when we view the rock as a whole, be considered exceedingly rare. There was, therefore, nothing which prevented the existence of animal life at the period in that area; and the fact of these remains being so found is highly satisfactory, though, should we not have detected them, there would be no proof that animal and vegetable life did not exist, even at no great distance from any spot where old red sandstone was discovered without them; for many circumstances may have prevented their preservation in the rock. We should, however, generally infer that deposits of sand or mud round coasts, or in shallow water, would contain, all other things being equal, more organic remains than those formed in deep water.

When we arrive at the carboniferous limestone itself, we have the first deposit of carbonate of lime of very great extent, and of a fair approach to purity, in the area now occupied by Western Europe; for though limestones are interspersed among the grauwacke rocks, and calcareous matter is often disseminated through them, cementing the grain of detrital matter which has been derived from more ancient rocks, the volume of limestone of equal purity in the one compared with that in the other is small. We do not mean to infer that this limestone is an uniform product of the period, even in the European area, much less that a contemporaneous deposit in America or China is necessarily and fundamentally a carbonate of lime; for such an inference would require perfectly equal conditions to prevail in all these situations at the same period. A considerable mass of carbonate of lime was, however, then thrown down in this part of Western Europe; and the abundance of organic remains detected in it, sometimes so great as nearly to constitute entire beds, shows that at least there was no scarcity of life in this part of the globe at that period. The great change in the nature of the deposit is remarkable, and shows the prevalence of the causes necessary to produce this effect for a very considerable period. Geologically considered, it may be regarded as only one among many periods; yet the time necessary to form this deposit, and to be sufficient to explain the facts connected with it, so little accords with the divisions of time which man commonly employs, that there is great difficulty in conceiving the amount of ages which may be included even in a single geological epoch.

The researches of Professor Sedgwick have shown that coal-beds, profitably worked, accompanied by sandstones and shales, occur at the same geological level, from Bewcastle Forest along the skirts of the

Cheviot Hills to the Valley of the Tweed, as the carboniferous limestone of the South of England and Wales. This great change by no means appears to be sudden, but gradual. A series of beds, intermediate between the mass of the carboniferous limestone and the coal-measures, and which in the southern parts of England is of little importance, swells out, as it were, in Derbyshire and Yorkshire, presenting subordinate beds of coal. Still proceeding northward, the carboniferous limestone itself becomes subdivided by coal, sandstone and shale, the latter finally acquiring considerable importance still further north, and the deposit taking the character of an ordinary coal formation, the calcareous matter having greatly disappeared*.

Now if it be fair to infer, as it appears to be, that the greater the abundance of coal with an abundance of vegetables in the associated beds, and which bear no marks of having suffered long transport, the greater the probability of dry land not having been far distant at the time of the deposit, we arrive at the conclusion that, at the period when the carboniferous limestone of the South of England was produced in the sea, there was probably dry land in the part of the European area not far to the northward of the present Tweed. We should further conclude that a gradual rise of this land was effected, by which means terrestrial vegetation travelled further to the south, so that its remains became abundantly entombed in that direction, producing coal now found in Southern England, Wales, and the line of deposits which appear

* Sedgwick, Address to the Geological Society, 1831.

in Belgium and Northern France, the continuity of the whole being superficially concealed beneath the red sandstone, oolitic, cretaceous and supracretaceous groups of a part of these countries.

This view would be borne out, if the freshwater deposit, with the remains of Cypris, Cytherina, and, as is supposed, of a saurian, discovered at Burdiehouse near Edinburgh, should turn out to be, as it is considered to be by Dr. Hibbert, subordinate to the carboniferous limestone; for the freshwater lake or river where the deposit was produced would necessarily require dry land. The discovery of saurian remains, if borne out by other circumstances than a tooth, would be most interesting, supposing no doubt can be entertained respecting the geological epoch to which these beds are referred, as it shows us that the condition of the atmosphere was not then unfavourable to their existence, though certainly we have no means of judging how far their respiratory organs were different from those of actual reptiles, enabling them to breathe an air more charged with carbonic acid. Mr. Murchison had previously pointed out that a freshwater limestone occurred at Pontesbury, Uffington, and other places in the coal-measures of the Shrewsbury district; and as this also would require dry land, we obtain another locality for it, which is the more interesting as the rocks being more recent than those of Burdiehouse, under the supposition that the latter are equivalent to the carboniferous limestone, it accords with the view we have taken of a gradual rise of land from North to South*. It cer-

* It has long been supposed that a large proportion of coal and its associated beds containing fossil plants were deposited in

tainly would by no means follow that such land should be unbroken; it might readily be divided into islands, in accordance with the views of MM. Sternberg, Boué, and Adolphe Brongniart. How far similar appearances would lead us to expect a continuous mass of this land in the direction of the present continent of Europe, at the epoch of the carboniferous limestone, it would be difficult to say; but it is highly interesting to observe that rocks equivalent to the millstone grit and limestone shales of English geologists become developed in Westphalia, and are there known as *Rauer sandstein*. These beds attain a considerable thickness at Arnsberg, Merscede and Warstein.

In some situations the coal-beds and associated strata of shales and sandstones, containing the remains of terrestrial plants, so alternate with limestones containing marine remains, that unequal conditions must have arisen during the deposit of the mass of the carboniferous group, and in many instances there must have been an alternation of such conditions. If violence had attended the transport of the plants now converted into coal or discovered fossil in the associated beds, the appearance of those in the latter would not be as we now find them: instead of appearing as if spread out by the botanist for examination, we should have had them crushed and disfigured. Moreover tranquillity seems requisite to explain the condition of those vertical or nearly vertical stems of plants discovered in the coal-measures of different situations* where they have been

fresh water, from the frequent absence of marine remains among them; but direct evidence derived from animal exuviæ of which the freshwater character could not be doubted, was wanting.

* Geological Manual, p. 406.

gradually enveloped by different beds of sandstone or shale, through which they therefore appear to pierce. These facts seem to require a slow depression of land beneath waters, in such situations that they were not exposed to the destructive attacks of waves or powerful streams of water; and if these vertical or nearly vertical stems were always so situated with regard to associated beds that we might presume the absence of the sea, large freshwater lakes, such as those of North America, would appear to be the localities offering the best conditions for their entombment.

The alternations of limestones containing marine remains, and of sandstones, shales and coal-beds with no trace of a marine creature in them, are exceedingly remarkable, and seem difficult of explanation without calling in the aid of oscillations of the solid surface of the earth, by which very gradual rises and depressions are effected. In these cases the views of Mr. Babbage respecting the expansions and contractions of rocks, caused by differences in the power of the surface to radiate heat, may eventually be found greatly to assist us, more particularly when coupled with central heat, which would cause such effects very generally at different geological periods. No doubt great contractions and slow dislocations of the earth's crust, caused by radiation of heat from the earth generally, would produce numerous similar effects from the variable manner in which the solid surface would be placed relatively to the waters resting on it; yet when we add the variations which might subsequently take place in the expansions and contractions of rocks, in consequence of the new position given to great masses, we obtain great additional aid in producing the effects required.

To trace even the probable positions of dry land over the European area at the carboniferous epoch would be most difficult, particularly when we recollect that what we term a geological epoch may include a long series of ages, and that thus land may rise and fall, be degraded and replaced, sometimes by the same, sometimes by greater, intensities of various forces than those the effects of which we daily witness, and yet the whole be included in a geological epoch, or rather one during which a particular group of rocks has been formed.

Although over a large part of England, more particularly the central and southern portions, the manner in which the red sandstone group rests upon the carboniferous group is such as to show that the latter was disturbed, dislocated, and partially removed before the former was accumulated upon it, there is great reason to suppose that in other parts of the European area deposits still continued quietly to be thrown down on undisturbed parts of the carboniferous series, so that no real line of separation can be well established between them. This, in fact, would be nothing more than that which took place relatively to the grauwacke and carboniferous groups, a portion of a given area being disturbed while other portions retained their original positions, deposits tranquilly taking place upon them. We should expect, if dry land continued long to exist at or near the undisturbed portions of a given area, that vegetable matter would be entombed in creeks, estuaries, deltas, and other situations. Now it is by no means improbable, making every allowance for the deceptive appearances caused by the coal-measure sandstones becoming red, as frequently happens, that the coal of some parts of Europe, particularly of Lower Silesia and Bohemia, may be equivalent to a part of the red sandstone series.

The prevalence of a particular kind of rock, such as the rothliegendes, the lowest member of the red sandstone group, over a considerable area, often remarkable for containing fragments of preexisting rocks of such size as to constitute a conglomerate, is certainly exceedingly striking, particularly when it is observed that fragments of porphyries are abundant in many places. In Devon the porphyry is found imbedded in the sandstones in fragments which even attain a size equal to three or four tons in weight. We will not here repeat what we have elsewhere* stated respecting the evidences of violent disruption in that part of England, further than to point out that the coast between Babbacombe and Dawlish affords excellent opportunities for studying it.

During the deposit of the rothliegendes, few organic remains were entombed in Western Europe. When we rise above this mass of red sandstones, marls, and conglomerate, which marks the great destruction of preexisting rocks over a large area, under circumstances which gave a general red tint to its deposit, we arrive, in the ascending order, at the second member of the red sandstone series, named the zechstein, or magnesian limestone. This is a calca-

* Geological Manual, p. 362.

reous deposit, of a somewhat variable aspect, though the extension of the lowest part of it, named the marl slate, or copper slate (*Kupferschiefer*), from Germany into England is remarkable, when we consider its small depth. This deposit is fossiliferous; and certain shells, *Productæ*, now appear in the ascending order for the last time, at least as far as observations have yet gone; and we arrive at evidence that a saurian, analogous to the present monitors, existed at or near those places where Mansfeld, Rothenburg-on-the-Saale, Glücksbrunn, &c., are now situated. We may therefore conclude that dry land was also then to be found in that part of the present European area. The copper slate is remarkable for containing numerous fishes of a particular genus (*Palæothrissum*, Blainville, *Palæoniscus*, Agassiz,) which are found in it as well in Northern England as in Germany. There is nothing in itself remarkable in finding the same kind of fish in two places not further distant from each other, more than there would be in finding herrings on the coasts both of Norway and France; but they belong to a family of fishes which, according to the researches of M. Agassiz, have not been detected above the red sandstone group. He states that the genera *Acanthodes*, *Catopterus*, *Amblypterus*, *Palæoniscus* and *Platysomus*, which form part of the family of *Lepidoides*, one that has no representative among known existing fish*, are not detected above this group.

The zechstein has not yet afforded any remains of trilobites, which were once supposed not to rise higher

* Agassiz, Recherches sur les Poissons Fossiles, tom. ii. p. 3. Neuchatel 1833.

in the series of rocks than the carboniferous limestone: the researches of Mr. Prestwich have, however, shown that they are found in the nodules which occur in the coal-measures of Coalbrook Dale. The species differ from those detected in the lower rocks; yet the fact of their being so found may lead us to suppose that, even among European rocks, there may be some chance of discovering the family continued up to the zechstein, a deposit which presents a general zoological resemblance to the inferior fossiliferous rocks, while it differs more considerably from those immediately above it.

The zechstein does not appear to be a deposit widely spread over the European area. As yet it is principally known in Germany and England. It merely appears that at a given time calcareous matter was thrown down in one place and not in another. We should expect to find organic remains in it, since the chances are that, when calcareous matter occurs as the zechstein does, we should discover organic remains more plentifully in it than in the other associated deposits, always excepting plants, which, viewed generally, are more abundantly detected in shales and sandstones than in limestones.

A series of beds of sandstone or marl covers the zechstein: it is of various tints of red, blue, white and green, thence known as the variegated sandstone (*grès bigarré, bunter sandstein*). It must not, however, be considered that there is anything peculiar in this respect to the rocks in question. Such appearances are frequent in various mechanical deposits, from the grauwacke upwards, and seem merely to depend upon the different states of oxidation of the iron, and, perhaps, sometimes of the manganese, in the different beds. They are therefore such as may be expected in the rocks of any age. The variegated sandstones do not abound in organic remains. Plants have been detected in them in Alsace and Lorraine, and it is remarkable that they differ from those in the coal-measures.

It is almost needless to observe when we contemplate the red sandstone series as a whole, and consider that it is in a great measure composed of matter which must have been deposited from water where it was, for the time, mechanically suspended, that great variations should be expected at the same geological levels; here clay or marl being found, there sandstone or conglomerate, while, occasionally, calcareous matter should be dispersed among it, under favourable circumstances, in sufficient abundance to constitute numerous beds of limestone. We should, therefore, expect considerable changes, in fact, precisely those which do occur, though often overlooked because they are merely from marl to sandstone, or from sandstone to conglomerate. When, however, limestone appears, the alteration in the character of the deposit is so obvious, and the rock itself employed for so many useful purposes, that this change is immediately observed, and we have it recorded. Moreover, as these accumulations of calcareous matter usually contain organic remains, they more readily attract attention, and therefore are seldom disregarded. The muschelkalk is a calcareous deposit of this kind, found over a more extensive area than the zechstein, but which does not extend into the British Isles, at least no trace of it has yet been detected in them, notwithstanding the great development of the red sandstone group in Great Britain.

The muschelkalk in the countries where it is found passes gradually into the rocks above and beneath: there is nothing to show any violent mechanical action; on the contrary, all tends to prove that the deposits proceeded tranquilly in those situations, and that the change was simply the substitution of one kind of matter for another. There was, it would appear, a considerable change in animal life at this period, at least as far as we can judge from the organic remains yet detected in the muschelkalk, which being merely negative evidence, is no doubt liable to much error. One remarkable circumstance connected with the zoological character of this rock is the absence of numerous species of corals, for as yet one only, *Astrea pediculata,* has been detected in it. Saurians now become more common, at least their remains have been more abundantly preserved; and we find the genera Crocodilus, Plesiosaurus and Ichthyosaurus, the former of which, supposing that its remains are certainly detected in the muschelkalk, has continued as a genus to inhabit the face of the globe since that early period until the present day, while the two latter have disappeared, apparently even before the deposition of the supracretaceous rocks, their remains not having been detected in the latter.

Above the muschelkalk there is an accumulation of marls mixed with subordinate sandstones, which have received the name of variegated marls, from the various alternating tints of blue, green, white, or red, which they present. The organic remains as yet detected in them have been principally discovered in

Wurtemburg, Alsace, Lorraine, the neighbourhood of Boll, and the countries situated in that part of Europe. Two genera of saurians, not yet discovered in the inferior rocks, Phytosaurus and Mastodonsaurus, are discovered in these beds. There is also the same Ichthyosaurus (*I. Lunevillensis,*) which is discovered in the muschelkalk, and a Plesiosaurus. We may conclude, both from these remains and the shells found in the same beds, and of which the analogous genera now inhabit shores and moderate depths, that land was not far distant from the situations where these remains are now discovered. This conclusion is supported by the remains of terrestrial plants found in the same districts.

It is worthy of remark, that as far as the European area is concerned, rock-salt is very frequently distributed among the higher parts of the red sandstone series, from which circumstance it has sometimes been named saliferous, by way of distinction; a term exceedingly objectionable, as it would imply either that at this geological period salt was more abundantly deposited than at any other, or that it was confined to it. Of the former, when we regard the whole superficies of the world, and not a minor part of it, such as Europe, we soon see that we have no proof whatever; and the latter we are certain is not correct, even when we examine the minor area itself: since salt is discovered in other rocks, even among the supracretaceous group. It is remarkable that salt, gypsum and dolomite, or rather a magnesian limestone, are so frequently associated with red or variegated marls or sandstones in different parts of the world. We know that many of these associations are of different geological epochs, and may infer that many others, supposed to be identical in distant countries, merely because their general mineralogical characters are similar, may also be distinct. There are many districts in various parts of the world where rocks with these characters prevail; and we may consider that the respective areas which they occupy have been placed under similar circumstances when the deposits were thrown down. Hence if we could approximate towards a knowledge of the causes which have produced the effects observable in one area, we might attain a better knowledge of the others, and be enabled to infer that a given part of the surface of our planet may have been under given circumstances at one, two, three, or more geological epochs, as the case may be.

The carboniferous and red sandstone groups of Europe have presented us with a series of detrital beds, in the lower parts of which a large amount of vegetation has been entombed. Local disturbances have here and there interrupted the quiet deposit of a continuous series of strata, so that we may expect to discover any higher portion of the series resting upon disrupted portions of a lower part. The greater proportion of the beds of which the whole is composed being apparently produced by the transporting action of water, we cannot suppose that the beds themselves can be continuous over very considerable distances: one must fine off and be replaced by another; so that, as has been previously observed, we may obtain contemporaneous equivalents which offer no mineralogical resemblance to each other, and which may differ in their organic contents, particularly at considerable distances, presenting an abundance of terrestrial remains in one place and of marine exuviæ in another, differences of conditions having necessarily produced different results. It will be obvious that it would be extremely unsafe to consider the coal of Australia, India, America and Europe to be precisely of the same age, merely because the sandstones and shales with which it is associated are generally similar. Even if some of the coal in each of these portions of the earth's surface were contemporaneous, as probably it may be, it would be equally unsafe to infer that there were not deposits, also contemporaneous, differing wholly from them, both mineralogically and as to organic contents, not a trace of a terrestrial plant being detected in them.

CHAPTER XVI.

We find above the red sandstone group an accumulation of sands, sandstones, marls, clays and limestones, named the oolitic group, because some of its limestones are of that kind called oolitic. It is a mere term of convenience, like those of 'carboniferous', 'red sandstone', &c., for many limestones in other groups are oolitic. A very considerable change must now have taken place over the European area. The waters wherein the detrital matter was held in mechanical suspension must have been much charged with carbonate of lime, so that very few of the deposits thrown down at this period are without calcareous matter. Marine animal life was exceedingly abundant over a large portion of the same area, so that some beds seem composed of little else than the remains of shells and corals. Viewed as a whole, there must have been great comparative tranquillity over a large portion of the European part of the earth's surface during the period of the oolitic group. Dry land must have existed in the same situations; for we find terrestrial plants which could not have been transported far, and even accumulations of them constituting beds of coal.

Reptiles were common, since their remains are so numerous; and the researches of botanists show that a perfectly different vegetation from that which preceded it flourished over the area under consideration.

There was, from all the facts brought to light, an entirely new set of plants. As far as these facts go, they show that "the proportional number of ferns was diminished, the gigantic Lycopodium-like and Cactoid plants of the coal-measures, Calamites, and Palms, all disappeared; vegetation had no longer a character of excessive luxuriance, but species, undoubtedly belonging to Cycadeæ, and analogous plants, now natives of the Cape of Good Hope and New Holland, appear to have been common. Coniferous plants were still plentiful, but they were of species that did not exist at an earlier period. Whether any other dicotyledons than those of the Cycas and Pine tribes existed at this time, does not appear*." A very striking zoological feature of this group is the immense abundance of Ammonites and Belemnites which must have existed previous to, and during, its deposit; for notwithstanding the usual chances of destruction to which we may suppose they were exposed, myriads of their shells have been entombed entire, and not unfrequently the animals must have been in them †. From all analogy, these creatures are supposed to have been free swimmers in the ocean: the seas occupying the area where the oolitic group

* Lindley and Hutton's Fossil Flora of Great Britain, Preface, vol. i.

† 173 species of Ammonites and 65 species of Belemnites have been enumerated as detected in the oolitic group. There is probably error as to the exact number of species of both these genera, shells in different stages of growth having been sometimes considered as distinct, and perhaps the latter having been unnecessarily multiplied from a desire to form new species; but still the number really distinct must be considerable. Numerous, however, as the species may be, the numbers of individuals found fossil in the oolitic group is still more remarkable.

is found must have swarmed with them. We do not through them attain a knowledge of the probable depth of water at the time; but as they are not unfrequently associated with marine genera, the analogues of which at the present day are inhabitants of shallow water or moderate depths, we have no reason to infer the existence of a deep sea, but on the contrary one of moderate depth.

The lias forms the lowest part of the oolitic group, and is remarkable for preserving a general character over a large part of Western Europe. In some places it appears to pass gradually into the group beneath it, while in others, as is the case in England, there has been a break in the deposit of matter at a given time. The detritus of which the lias is composed must have been extremely comminuted, and have been mixed with calcareous matter at particular times, more especially in its lower portions, on a surface extending over a considerable part of Europe. Among the reptiles which seem to have abounded at the time of the lias, we detect some, such as the Pterodactyli, Plesiosauri, and Crocodiles, which must have required dry land and creeks, bays and other sheltered places. Plants, frequently well preserved, appear to show that land was not far distant. We therefore endeavour to discover the situations where we may consider such land to have existed. Here, however, our data fail us. We may indeed assume that the grauwacke of Normandy and Brittany rose out of the waters of the period, since the oolitic rocks quietly cover up their tilted strata, filling preexisting inequalities; and we might infer the same with other districts. The European area has, however, been so broken up by dislocating causes, and so often abraded by the action of water since the period of the lias, and indeed of the whole oolitic group, that we can make little advance in inquiries of this kind.

The saurian remains of the lias, both in England and Germany, very frequently present little appearance of having lain long exposed to the effects of decomposition before they were entombed; on the contrary, many skeletons of Ichthyosauri show, by the contents of their stomachs still remaining between their ribs, while there are even traces of skin upon their bones, that their death must have been speedily followed by their envelopment in the detrital matter of the lias, if, indeed, they were not often entombed alive. The lias frequently contains a large quantity of the excrements of the various creatures which existed at the time; and it not unfrequently happens that there are lines of these coprolites at various levels in different localities, as if the muddy bottom of the sea received small sudden accessions of matter from time to time, covering up the coprolites and other exuviæ which accumulated during the intervals.

It would far exceed our limits to enter upon the inferences which might be deduced from the various facts that are known respecting the oolitic group; we must therefore content ourselves with a general glance at it as a whole. The lias of Western Europe may be considered as a deposit of finely comminuted detritus, in which, though calcareous matter is not rare, it was only occasionally sufficient to produce limestones. The sands of the inferior oolite, into which the lias passes upwards, range over a considerable area, extending from Northern far into Southern Europe; the causes, therefore, which have produced them have been equally general, and their mineralogical character shows a general change of the bottom of the sea upon which they were accumulated: instead of being mud, it gradually became arenaceous, and finally calcareous, in the beds known as the inferior oolite itself. Calcareous matter is, indeed, not wanting in the sands, but we are not certain how much of it was present in the actual deposit, and how much may have been added to it by the subsequent percolation of calcareous matter downwards from the limestones of the oolite. In many places, extending from England to the Jura chain, grains or small nodules of hydrate or oxide of iron are common in the inferior oolite, so much so as to afford iron for the foundries in the department of the Haute Saone.

The next portion of the oolitic group is highly interesting, as its organic contents point to the proximity of dry land at that period over a part of the European area extending from Northern England and Scotland into Germany. Terrestrial plants abound in these situations, the accumulations of them being frequently sufficient to produce beds of coal. Among them Mr. Murchison considers he has observed vertical stems of *Equisetum columnare* in sufficient abundance and over a sufficient area to justify him in concluding that a considerable area gradually sunk so quietly, that these plants became tranquilly imbedded in the matter which accumulated around them. When we examine the beds which are apparently equivalent to these in the southern parts of England, we find marine remains, and an abundance of calcareous matter, the various strata into which they are separated being grouped by geologists into smaller portions, known in the ascending order by the names of fuller's earth, great or Bath oolite, Bradford clay, forest marble and cornbrash. Now it is a highly interesting fact, that a remarkable accumulation of organic exuviæ at Stonesfield, among which are the remains of the only mammiferous creature yet detected beneath the supracretaceous group, has been found by Mr. Lonsdale to be a lower portion of the great oolite. These remains have been considered referrible to those of a Didelphis, or of an analogous quadruped, and to have belonged to two species. With them we find the remains of terrestrial plants, of marine shells, of a huge saurian, the *Megalosaurus Bucklandi*, and bones which probably belonged to a winged terrestrial reptile, the Pterodactylus. We may therefore infer that immediately succeeding the period of the inferior oolite, and while the great oolite of Southern England was forming, dry land extended towards the area now occupied by the central part of our island. Such land may indeed have constituted islands, and eventually there may have been extensive coral reefs, since polypifers are common in the upper part of the great oolite itself, both in Southern England and in Normandy.

After this period a larger part of the oolitic area was probably again submerged; for we find a mass of clay, continuous over considerable distances, and containing marine exuviæ, covering over the terrestrial remains, which were so common in Northern Britain and a part of Germany. The remains of saurians, among which are those of crocodiles, being however often detected in it at different points, we may still suppose the existence of dry land, in various parts of the area, though changes in the relative situations of land and water had produced such physical differences in parts of it, that mud was commonly deposited, while quantities of terrestrial plants were no longer accumulated in particular situations. Arenaceous matter was, however, again thrown down over a large part of the area, and was succeeded by a calcareous deposit, the former known as (the lower) calcareous grit, and the latter by the name of coral rag, from the abundance of polypifers detected in it. Coral reefs abounded over the area once covered by deep mud, arenaceous matter being first accumulated upon the latter. This change is sufficiently remarkable, but is rendered still more so by being similar to other changes which succeed, the calcareous rocks being first covered by an arenaceous deposit (the upper calcareous grit), to which another accumulation of mud (the Kimmeridge clay) succeeds: this is again covered by sands (the Portland or Kimmeridge sands), and the whole group is crowned by limestones (the Portland oolite).

We are not to suppose such changes in the bottom of a sea occupying a given area to be constant over the whole area; but the extent to which each of them has been traced is exceedingly remarkable. The clay beds can be followed, as might be expected from the fine condition of the detritus from which they have been produced, with similar general characters, over larger areas than the sands or sandstones; and the calcareous deposits, as far as regards similarity of structure, over smaller areas than the sands. There are necessarily numerous minor changes resulting from the operation of minor causes; but the general character of the whole oolitic group over a large part of Western Europe is remarkable.

There can be little doubt that this group, greatly expanded in thickness, and mixed with sandstones, marls and slates, possessing a very different aspect from the equivalent rocks in a large portion of Western Europe, extends over various parts of Eastern Europe. There can also be little doubt that it constitutes a large part of the calcareous Alps, extending into Italy and the countries of South-eastern Europe. Its mineralogical appearance is more that of the grauwacke series, showing that there have been different modifying circumstances, producing different though contemporaneous results, in different parts of the whole European area. While an abundance of organic remains characterizes the western portion, fewer organic exuviæ are found in the eastern part. We may readily suppose shallow water, perhaps interspersed with land, in the one situation, while deeper water and less land occurred in the other.

We must not omit to notice that singular collection of organic remains in certain calcareous rocks, known as lithographic slates, and which occur at Pappenheim, Solenhofen, and Monheim near Eichstadt. They are considered to constitute the highest part of the oolitic series of Germany. In these beds are entombed the remains of six species of that extraordinary flying reptile the Pterodactylus, and with them the exuviæ of numerous insects, (Libellula, Æschna, Agrion, Myrmeleon? Sirax? and Solpaga?) which probably constituted their prey. Among the reptiles we have also the remains of the Geosaurus, Lacerta, Rhachcosaurus,

Ælodon (Crocodilus, *Jäger*,) and the Pleurosaurus. We may hence assume the proximity of dry land. The crustacea found in these beds consist of Pagurus, Eryon, Scyllarus, Palæmon and Astacus, probably frequenting coasts. There are several molluscs, and among them one named Aptychus (*Meyer*) of a singular form. We may readily suppose the whole deposit to have been effected on a coast where the water was not deep, on the shores of which the Pterodactylus chased its insect prey. The plants noticed in the deposit are marine, which may easily have been entombed, as Fuci now are, in similar situations. When we reflect that there is evidence of the existence of dry land in Southern England, immediately succeeding the deposit of the Portland oolite, and that a considerable amount of sands and clays, known as the Wealden Rocks, show that such dry land continued to exist for a considerable period in the same part of the European area, the Solenhofen beds acquire additional interest; since, though they may not be precisely contemporaneous with the Wealden deposits, they show dry land at nearly the same epoch in the place now constituting part of Bavaria, and in that now occupied by South-eastern England.

It is only necessary to take a common artificial globe in our hands to see the small portion of its superficies representing the whole area of which we have been treating. Contemporaneous rocks were no doubt produced in numerous other parts of the earth's surface; and there is reason to conclude that rocks with somewhat similar organic remains have been found connected with the range of Himalayan mountains. Time must necessarily elapse, and observations be greatly multiplied, before any large portion of the earth's surface can be brought, as it were, under the immediate view of the geologist,—even before he can estimate the probable condition of a fiftieth part of such surface at any given period. There is, however, evidence in the area treated of to show a considerable change, both in the nature of the deposit thrown down at the epoch of the oolitic group, and in the organic remains detected in it, from the condition of the same area at the time of the red sandstone series. Carbonate of lime greatly more abounds in the one than in the other; indeed the spread of this substance over a large comparative area at this period is a remarkable fact. How far other parts of the northern hemisphere may present us with equivalent rocks, now above the level of the sea, future observations must determine; and it will be exceedingly interesting to observe, as facts become multiplied, how far the general organic structure of life may have been similar at this period, or how far advances may have been made to that diversity of distribution which we now observe, assuming that at the earlier fossiliferous epochs there was a greater approach to uniformity. At present it seems to be considered that rocks equivalent to the oolitic group have not been detected in North America. There is, however, so much of that great area of dry land still unexplored, that no inferences can fairly be drawn from the facts yet known on this head.

There is evidence immediately above the oolitic group in Buckinghamshire, the Vale of Wardour, the vicinity of Weymouth, and in the Boulonnois, that dry land existed in such situations, and that trees and plants analogous to those now growing in warm climates flourished upon it. In the vicinity of Weymouth these plants remain rooted in, or fallen upon, the soil on which they grew, though covered up by more recent deposits. The soil thus preserved is named the *dirt-bed*: in it, or rather them, for there appear to be two or three of these *dirt-beds*, we occasionally detect rounded pebbles of the inferior rock. There is, however, no evidence of the violent action of water generally on the surface of the Portland beds on which they rest, so that we may consider the pebbles as resulting from causes similar to those constantly in action on shores and in rivers. As the whole is based on a rock replete with marine remains, an elevation of this rock, and in this place, must have preceded the growth of terrestrial plants upon it.

Above this dry land a series of deposits has been thrown down. characterized by the presence of terrestrial, freshwater, or estuary remains; and as these rocks are well developed in the Weald of Sussex, they have been named the *Wealden Rocks.* The exact area occupied by the deposits thus characterized cannot be ascertained, since they have been covered by the cretaceous series, the removal of which by denuding causes has permitted us to know that they have existed. The great extent of these rocks observable in the Weald of Sussex has been brought to light by an elevation of land in that point, and by a subsequent removal of the superincumbent cretaceous series. As far as we can perceive, the south-eastern part of England appears to have been the deepest portion of an estuary, into which detrital matter was drifted, and entombed the organic remains above noticed. We have seen that in the vicinity of Weymouth dry land existed immediately preceding the deposit of the Wealden rocks; and it does not appear, from the facts there observable, that these rocks extended much to the westward of that point, since the cretaceous series quietly overlaps the various members of the oolitic group in a western direction.

The Wealden rocks are composed, in the ascending order, of various limestones, alternating with marls, named the *Purbeck beds*; of numerous strata of sands and sandstones, often ferruginous, and occasionally intermingled with shales, named the *Hastings sands*, from being well developed at that place; and of a clay known by the name of the *Weald clay*. We are indebted to Dr. Fitton for pointing out the true nature of this remarkable series of beds, and thus showing that there must have been an extensive estuary, or some analogous collection of brackish or fresh waters, at this given epoch in a particular portion of the European area. The organic contents of these beds, though not numerous as regards species, are highly interesting. We learn from the labours of Mr. Mantell that a huge reptile, more resembling the Iguana in its osteology, particularly in its teeth, than any known creature of the present day, (and thence named the *Iguanodon*,) roamed on the shores of this lake or estuary, living probably on the plants now found fossil with it. Its reptile companions were no less remarkable; for the estuary and its banks were tenanted by the Hylæosaurus (another fossil reptile brought to light by the labours of Mr. Mantell), the Megalosaurus and the Plesiosaurus, all genera which no longer exist on the face of the earth. We have seen that the two latter were entombed in rocks previously deposited, but the genera Iguanodon and Hylæosaurus now appear for the first time, as far at least as researches have yet gone. Being apparently terrestrial creatures, and therefore the chances of the preservation of their bones in rocks being far less than if they were marine, it cannot be stated that they were now first created, since we are considering phænomena observable on a mere point of the earth's surface, and since a combination of circumstances is required not only to preserve their remains, but also to expose such remains, when preserved, in situations where they can be examined. With these extinct genera we find the exuviæ of the Crocodile, Trionyx, Emys and Chelonia, so that there was no want of reptiles in the district in which the deposits under consideration have been produced.

If we consider these rocks to have been formed in an estuary, and not in an isolated mass of water, we should suppose that there must have been a considerable tract of dry land over the surface now occupied by Southern England and Northern France at this period. We have observed that the trees in the *dirt-bed* in the neighbourhood of Weymouth stand in, or rest upon, the soil in which they grew; consequently their submersion beneath the water in which the Purbeck beds were formed, and which now cover them up, must have been gradual and unaccompanied by a rush of waters. Had there been a violent wash of waters, or had the land been long exposed to the destructive action of breakers, the trees and the soil would have been speedily removed. Hence not only must the submersion have been gradual, but the soil also could not have been exposed to any considerable surface of water, which, either fresh or salt, would have rolled in breakers on the coast, unless indeed it were protected by trees analogous to the mangroves of the present tropics.

We have reason to suppose that the land was gradually depressed in this part of the European area in such a manner that the resulting cavity was occupied by fresh water, and that, either from differences in level or other causes, the sea did not flow freely into it. Much calcareous matter was first deposited, and in it were entombed myriads of shells, apparently analogous to those of the Vivipara. Then came a thick envelope of sand, sometimes interstratified with mud, and, finally, muddy matter prevailed. The solid surface beneath the waters would appear to have suffered a long-continued and gradual depression, which was as gradually filled, or nearly so, with transported matter: in the end, however, after a depression of several hundred feet, the sea again entered upon the area, not suddenly and violently, for the Wealden rocks pass gradually into the superincumbent cretaceous series, but so quietly that the mud containing the remains of terrestrial and freshwater creatures was tranquilly covered up by sands replete with marine exuviæ. The depression did not, however, stop here, or rather the differences in the levels of the sea and land in this particular part of the earth's surface continued to increase in such a manner, that the bed of the estuary or lake, in which the Wealden rocks were formed, became depressed at least 300 fathoms beneath the level of the sea, and, as it would appear, in a most gradual manner, for there are no marks of violence in the cretaceous rocks that repose upon the Wealden deposits.

As estuaries of the kind above noticed must necessarily occupy a comparatively limited area, the effects produced must, to a certain extent, be local, if even we suppose that the estuary character of the deposits extended over a surface equal to any modern river-delta. Contemporaneous littoral deposits which should form continuous portions with the sand, mud, or silt brought down by the river, would, we should conceive, be full of marine remains, precisely as on the coasts of Africa or America, at the present day, long lines of coast must contain, in the same continuous surface of soundings, either marine or estuary remains, according as large rivers do or do not pour their waters into the ocean which bathes the coast. Where, therefore, circumstances were favourable, we should expect to discover marine equivalents of the Wealden rocks, which from all appearances must have required a long lapse of time for their production. It has been considered, from the freshwater character and geological position of certain beds in the Pays de Bray, near Beauvais, where a denudation of the cretaceous series exposes the inferior rocks, that these deposits are equivalent to the Wealden deposits. This may be perfectly true, and yet the deposits not be continuous portions of detrital matter thrown down in the same delta or estuary. Perhaps we shall never possess sufficient evidence to show that they either did or did not constitute different parts of the same delta or estuary; but if they did, we should, from analogy, infer that a large surface occupied by dry land was necessary to the production of a river equal to the Quorra or the Ganges.

Certain rocks with marine remains occur in different parts of Europe, in France, Switzerland, Germany and Poland, which occupy a situation intermediate, as it were, between the cretaceous and oolitic groups, and therefore equivalent to the Wealden rocks. It is evident that there must have been marine equivalents of these rocks, and from all analogy we may conclude that such marine equivalents were greatly predominant even in the European area, itself of trifling extent compared with the superficies of the globe.

Above these Wealden deposits we find the cretaceous series, which from its extent must have resulted from causes not only common to the European, but also to at least a portion of the Asiatic area, since it is evidently found in that part of the world. Viewed generally over a considerable portion of the surface covered by it, the well known rock, called chalk, is based either upon an arenaceous or argillaceous deposit, the mechanical origin of which can scarcely be doubted. As we cannot expect uniformity in the distribution of detritus, unless under conditions which could scarcely ever obtain, we should anticipate that sands would predominate over one part of the area, and mud or silt over another, that there would occasionally be alternations of these substances, and that frequently particular deposits of sands or mud would be found continuous over comparatively extensive surfaces. It is not to be expected that the divisions of the lower part of the series into upper green-sand, gault, and lower green-sand can be traced to great distances from the southern portions of Great Britain, where these distinctions have been found so useful. As, however, their continuity with similar characters is remarkable in England, and shows the uniform operation of given causes somewhat extensively, we may anticipate that, with due attention and allowances for local variations, analogous divisions may be detected over some parts of the neighbouring continent, as, indeed, appears to be the case in part of Northern France and in its continuation into the cretaceous district of Aix-la-Chapelle.

The upper portion of the series is exceedingly remarkable for the constancy of its mineralogical characters over a surface extending from Northern France, through the British Islands, Northern Germany, Denmark and Sweden, into both European and Asiatic Russia. Now although equivalents of the cretaceous rocks are abundant in other parts of Europe, to the southward and eastward of this range of white chalk, they generally possess a somewhat different character, compact grey and dark-coloured limestones and compact sandstones being apparently of the same geological age as these white chalk rocks. This difference necessarily points to a modification of origin, so that we should anticipate some circumstances, productive of this remarkable white calcareous rock, to have existed over a considerable curved portion of surface not common to the whole European area.

We can scarcely consider that the white chalk was produced along a line of coast to which nothing but this calcareous matter, mixed with silica, was borne down; since this would suppose a state of things so utterly unlike that at present existing, or which we should imagine ever likely to have existed, that we seem compelled to seek some other explanation for the phænomena observed. The white chalk has more the appearance of a chemical deposit, or rather one resulting from a quick precipitation of carbonate of lime from solution in water, by expelling an excess of carbonic acid which kept the carbonate of lime in solution. If the waters were suddenly heated, a quick precipitate would ensue, and thus we might obtain the friable carbonate of lime. This is, however, not the only condition required. There is an abundance of silica, which, from the mode in which it is deposited, we should infer had been in solution. Now the abundance of silica which has separated out from the chalk generally is very considerable, so that we require an abundant solution of silica as well as of carbonate of lime. We must also not forget that numerous marine creatures existed at the time, so that if the mass of chalk were deposited slowly, the conditions necessary for the support of animal life also existed. The question would therefore arise, Has the mass of white chalk been slowly or somewhat suddenly produced?

Organic remains are in general beautifully preserved in the chalk: substances of no greater solidity than common sponges retain their forms, delicate shells remain unbroken, fish even are frequently not flattened*, and altogether we have appearances which

* The author remembers to have seen some specimens of fish from the Sussex chalk exhibited to the Geological Society by Mr. Mantell which were particularly remarkable in this respect, the fins projecting as if they were swimming in the water; their appearance being thus so very different from the fossil fish usually discovered, the fins of which are almost always compressed. The author was particularly struck with the fact at the time, as it seemed to show the kind of light precipitate in which they appear to have been entangled prior to their death. These beautiful specimens are probably still in Mr. Mantell's collection.

justify us in concluding that, since these organic exuviæ were entombed, they have been protected from the effects of pressure by the consolidation of the rock around them, and that they have been very tranquilly enveloped by exceedingly fine matter, such as we should consider would result from a chemical precipitate. How far the white chalk may have been the result of sudden precipitation, or of a succession of sudden precipitations, it would be difficult to say; but that the carbonate of lime itself fell gently upon the matter beneath it, and speedily became consolidated, seems proved by the condition of the organic remains found in the chalk.

Let, however, the origin of the white chalk have been what it may, the conditions for its production did not, as above observed, extend over the whole area occupied by the cretaceous series. It is not to be supposed that a large portion of the earth's surface would be covered by this fine precipitate of carbonate of lime at the same time; on the contrary, we should expect to discover rocks of equal age, formed in the ordinary way from the deposit of detritus, entombing organic remains in the common favourable situations. Where rocks have not been continuously traced, it is always difficult to be certain that we have an exact equivalent of a given part of a series at distant places. Various modifying causes would obviously produce differences difficult to determine. According to M. Dufrénoy, the salt of the celebrated Cardona mine occurs in the upper part of this series, as is also the case with the rock-salt of Mon Real. At Angoulême the inferior arenaceous rocks of the series are crowned by a limestone almost saccharine. Again, in the Alps,

Apennines, and the countries extending from Greece eastward into Asia, the cretaceous rocks assume a character different from those of Western Europe.

A great abundance of organic remains has been detected in the cretaceous series, amounting to about 160 genera, and 768 species, and it has been found that the latter, with a few doubtful exceptions, differ from those detected in the inferior groups, over those areas at least where they have yet been examined. Terrestrial plants are distributed, though perhaps not very abundantly, among the inferior sands and clays of Western Europe, many pieces of wood showing by their condition and the holes drilled in them by the Teredo, or analogous creatures, that they have been long drifted on the ocean. In the white chalk, however, marine plants are detected, and terrestrial plants must be very rare, since their presence has been much questioned; though why we should not expect to find them does not appear clear. Their general absence, from a large part of the white chalk area, seems, however, to show that dry land whence plants may have been drifted was not common near such portions of the white chalk. Hence we may conclude that at the epoch of the cretaceous series there was a very general difference, over a considerable part of the European area, in the condition of such part of the earth's surface from that presented at the epoch which immediately preceded it. There was, however, an accumulation of vegetable matter, sufficient to produce coal, even so near the great mass of rocks in which plants are scarce as Pereilles near Bellesta, Ernani near Irun, and at St. Lon in the Landes, where, according to M. Dufrénoy, coal is found in this series. Coal is also stated to occur in a cretaceous sandstone near Quedlinburg.

As yet no remains of mammiferous animals have been discovered in the cretaceous series; and the exuviæ of reptiles are by no means common, presenting in this respect a marked difference in its zoological character from that of the oolitic group which immediately preceded it. Hence we might infer that dry land was not so abundant in the European area at the cretaceous, as at the oolitic epoch. The remains of a crocodile have been found in the chalk of Meudon, and the exuviæ of a large reptile, the Mososaurus, at Maestricht and in Sussex. The remains of fish are not so rare, and it may even be said that fish-teeth, apparently those of sharks, are somewhat common. Molluscs were entombed abundantly at this epoch. Ammonites, though still frequently found, do not appear to have been so common as at the oolitic period, when they must have swarmed in the seas now replaced by the dry land of Europe. Other camerated shells, however, Scaphites, Hamites, Turrilites and Baculites, were now comparatively abundant. Indeed at one time these genera were supposed to be peculiar to the cretaceous series; but since Hamites and Scaphites have been detected, though very rarely, in the oolitic group, we may anticipate that the other genera may hereafter be observed in other rocks, particularly in distant parts of the world. From analogy we should expect that no very considerable depth of water was required for the various molluscous genera found fossil in the cretaceous rocks. The Echinite family may perhaps be considered to have been abundantly distributed at this epoch, more particularly during the production of the white chalk. Corals, generally speaking, do not seem to have prevailed in the seas to the same extent as at previous periods, while the spongiform zoophytes seem to have been more abundant. We should, however, be somewhat cautious in the latter inference, since so much must depend on those circumstances which could preserve bodies so easily destroyed, and which appear to have been highly favourable at the period of the white chalk.

Like all other deposits which require considerable time for their production, we should expect to discover rocks, equivalent to the series under consideration, in various parts of the world, where circumstances had been so far favourable that they have been subsequently elevated above the surface of the waters in which they were formed. Dr. Morton has described a series of fossiliferous beds in North America, which he infers, from their organic contents, were produced at the same time with the cretaceous rocks of Europe. The evidence on this head, if we take organic remains as our guides, would certainly lead to this inference, and therefore we may consider the American ferruginous sand formation (the name given to this deposit) as equivalent to the cretaceous series of Europe until evidence be adduced to the contrary, though, from the facts known respecting it, we should rather anticipate that further research would confirm the inference, since its connexion with superior rocks points to considerable general analogy between European and American deposits of this particular date. Assuming that there can be no doubt of the determination of the saurian remains detected in the American deposit, we find that the Mososaurus existed in that part of the earth's surface, at least as a genus, at the same time that it lived in that portion now occupied by Europe, and that it was also accompanied by Crocodiles, the remains of which are stated to be abundant. It is interesting also to observe that the Plesiosaurus and Geosaurus (saurians the remains of which have not been yet discovered so high in the series as the cretaceous rocks of Europe,) are associated with the exuviæ of the Mososaurus in America. Another creature, the Saurodon, found in the ferruginous sand of America, has not yet been noticed in any European rock.

A considerable portion of dry land was probably submerged in the European portion of the earth's surface to permit the deposition of the cretaceous series, which seems to have overlapped a large area, occupied by a great variety of preexisting rocks, from the gneiss and mica-slates of Northern Europe to the oolitic and Wealden deposits of more southern countries inclusive. This condition of the European area was destined to suffer a great change; the chalk was brought within the destructive action of waters; its surface over a considerable part of Western Europe was eroded; and the harder portions, consisting of flints, for the most part remained on its eroded surface, frequently showing, by their angular condition, that they have neither suffered transport to considerable distances, nor any great amount of attrition during either long or short periods of time. There must have been a considerable rise in the mass of chalk, and probably of at least a large portion of the European area, relatively to the level of the sea, so that dry land existed, in places which had previously been occupied by sea, supporting vegetation, and eventually an abundance of terrestrial animal life.

CHAPTER XVII.

THE supracretaceous deposits are very commonly termed Tertiary. This is, however, a name exceedingly objectionable, as it would imply that there were three great classes of rocks possessing marked characteristic distinctions, and that the deposits above the chalk constituted the third of such classes. When MM. Cuvier and Brongniart communicated to the scientific world their valuable labours on the rocks above the chalk round Paris, the chalk itself was considered the highest known rock, and the supracretaceous deposits as mere superficial sands, gravels, or clays. Hence, when it was found that the Parisian beds did not pass into the chalk beneath, but that the surface of the latter had evidently been abraded prior to the deposition of the former, and that the organic contents of the supracretaceous rocks differed so remarkably from those of the chalk, it was natural to impose the term tertiary upon the supracretaceous deposits of the neighbourhood of Paris, as rocks were then divided into two great classes, the Primitive and Secondary, between which was a kind of ambiguous class known by the name of Transition, the chalk constituting the highest of all the rocks then admitted by geologists. It is not, however, to be supposed that a great break took place at this time in the deposits over the whole surface of our planet, neither should we anticipate that even in the European area itself we should everywhere detect a solution in the continuity of deposits formed at this epoch. We should rather consider that as other groups of rocks pass into each other in one place, while there is evidently a geological break between them in another, that the like should happen with respect to the cretaceous and supracretaceous groups. The grauwacke is separated from the carboniferous group in Northern England in a manner which shows far greater violence than any facts, yet brought to light, would lead us to suspect had occurred at the time intervening between the formation of the cretaceous and supracretaceous rocks of Western Europe, and yet we see that in Southern England the grauwacke gradually passes into the carboniferous group. *A priori*, therefore, we should anticipate that, in the European area itself, we should detect a passage of the cretaceous into the supracretaceous series. Now Professor Sedgwick and Mr. Murchison consider that they have detected such passages in certain rocks of the Alps, at Gosau and other places; and the researches of M. Dufrénoy would lead to the same inference. At Maestricht there are evidently beds above the chalk not commonly found in the cretaceous districts; somewhat similar deposits also occur in Normandy. At Maestricht, however, the upper surface has been abraded prior to the production of the superincumbent beds, so that they do not pass into each other. But as we do not know the amount of matter removed by denudation prior to the deposit of the superincumbent beds, we are by no means certain that there may not be an absence of numerous beds, which once existed, and which have been swept away. We only learn the probable existence of many other rocks, over areas no longer extensively occupied by them, from patches which have here and there remained, and resisted the denuding forces that have carried away many cubic miles of solid matter from given areas. A considerable change no doubt was effected over a large part of the surface now occupied by Europe, but when we take a common artificial globe into our hands and see what that area is, we readily perceive the insufficiency of the data on which it is attempted to found a grand system of rocks applicable to the whole surface of the earth.

The supracretaceous rocks constitute a large portion of the dry land of Europe; and numerous memoirs have been written to show the probable amount of surface occupied by various lakes and seas in which numerous terrestrial, fluviatile, lacustrine and marine remains have been imbedded. To enter upon the probabilities of the existence of the various minor areas, noticed as either occupied by water or dry land, as the case might be, would far exceed our limits. There is evidence to show that numberless changes requiring a great lapse of time have taken place between the deposit of the chalk and the present order of things, and that as we approach the latter, there has been a general increase of dry land on this part of the earth's surface.

Among the various complicated changes which have taken place at the supracretaceous epoch, it has been considered the better plan to arrange the rocks referrible to it according to their organic contents, those deposits being considered the most modern which show the greatest approach, in the organic remains found in them, to the existing animal and vegetable life of the countries where they may be situated.

Though it has long been the opinion of geologists that the newer the deposit the greater was the probability of detecting the remains of existing animals and plants in such deposit, yet no general divisions of the supracretaceous rocks founded on this principle were until lately attempted for the European rocks of this age. Indeed it was long supposed that all supracretaceous rocks were referrible to some one or other of those found in the vicinity of Paris. The first general advance beyond this view appears to have been made by M. Desnoyers*, who pointed out that the supracretaceous deposits of Paris, London, and the Isle of Wight were among the more ancient of these rocks; that several divisions of the supracretaceous strata might be established, arising from oscillations of the land and the consequent deposit of matter in seas and rivers; and that all these periods presented, in their deposits and in their fossils, a progressive and insensible passage from one to the other, from the ancient state of nature to the present, from the more ancient supracretaceous basins to the actual basins of our seas. M. Deshayes, having examined a multitude of the shells of this period, collected from different situations, divided the whole series into three sub-groups according to their organic contents. The tables formed by M. Deshayes were published by Mr. Lyell, who informs us that he had, without a knowledge of M. Deshayes's classification, divided the supracretaceous group into four parts †. In order, however, to render the tables available, he has only given three distinct names to the parts of the group, M. Deshayes's upper division being subdivided

* Annales des Sci. Nat., 1829.

† Principles of Geology, vol. iii., Preface, 1833.

into newer and older Pliocene, the remaining two being called Miocene and Eocene. These names are expressive of a gradual approach of the fossil molluscs contained in the rocks under consideration to those which live in the present seas, few analogues of existing species being detected in the Eocene division, while they greatly prevail in the Pliocene. These distinctions are therefore essentially founded on change, and can in no way support the opinion that nothing but repetitions of similar events have succeeded each other after animal and vegetable life first existed on our planet, since they imply a gradual disappearance and nearly total destruction of given molluscs over an area which subsequently became covered with others that have never been discovered in any older deposits, and which therefore we conclude have been created to replace those that have disappeared as living creatures from the surface of the earth.

Classifications entirely founded on organic remains are at all times liable to be erroneous, if contemporaneous deposition be thence inferred as a necessary consequence, as we have had occasion to observe; they therefore may be considered as doubly liable to error when employed in proving contemporaneous origin in such rocks as those of the supracretaceous period, and which may contain a certain per centage of the remains of molluscs resembling those of the present day. Classifications of this kind necessarily infer a series of equal changes and conditions at numerous distinct and often distant places at the same time. When we appropriate names of this kind to rocks derived from one character alone, and that character one which cannot be considered constant, more particularly in the deposits under consideration, we theoretically prejudge the place which the rocks should occupy in a geological series, where the comparative date of the deposit itself is sought, not the particular proportion of given organic remains detected in it, which is necessarily a secondary consideration. If this were the course generally adopted, contemporaneous deposits of other rocks, which may contain different proportions of given organic remains, would be separated even in the area of England. To seek the different conditions under which animal and vegetable life were entombed at a given geological epoch, would evidently be impossible if we are merely to collect a number of fossil molluscs, place them before the conchologist, and then arrange our deposits theoretically, according to the per centage of existing or non-existing shells he may find among them.

We should probably find few geologists disposed to contend that because supracretaceous rocks may be detected in India, America and Europe, each containing about 50 per cent. of existing molluscs, they were therefore precisely contemporaneous, and that they should be classed under a particular head which implied that equal conditions obtained in these various parts of the earth's surface at the same time. If it be considered convenient to divide the supracretaceous rocks of Europe into three or more sub-groups, names which imply their actual geological position in the series, such as 'superior,' 'medial' and 'inferior,' 'upper,' 'medial,' and 'lower,' or others of the like kind, would seem preferable to those derived only from a per centage of certain organic contents, which seems to throw the determination of the place any given fossiliferous rock should occupy in the series of deposits generally, not into the hands of those who have carefully studied all its varied relations in the field, but into those of the conchologist in his cabinet, who probably never saw a square yard of the country where the particular deposit was situated. We are very far from undervaluing the aid of experienced conchologists, who have so greatly contributed to the advance of geology; we are only unwilling to see the study of the fossiliferous rocks turned into a mere cabinet examination of organic remains, to the utter neglect of all the other circumstances connected with them, which are frequently of far more importance *.

It seems very generally agreed that the supracretaceous rocks of what have been termed the basins of Paris, London, and the Isle of Wight, are among the lowest known in Western Europe. Those round Paris have been long celebrated for the extinct animals brought to light by the brilliant researches of Cuvier, and for the joint labours of the latter author and Brongniart, which showed that the Parisian area had been exposed to several changes, by which marine and freshwater and terrestrial creatures had been alternately entombed in the deposits then formed.

* In thus endeavouring to show that the terms Pliocene, Miocene and Eocene belong to a class of names which necessarily prejudge an important part of geological inquiry, and therefore should not be employed, though it is extremely probable they will be, we are very far from wishing to detract in the slightest degree from the value of the labours of their author, Mr. Lyell, among the rocks under consideration, believing as we do that they are highly important, and that they will eventually lead, notwithstanding the particular theory with which they are coupled, to a more complete knowledge of the supracretaceous deposits of Europe than has yet been attained.

Dry land necessarily first existed in that neighbourhood, since the lowest supracretaceous rock, named the plastic clay, seems to have been accumulated in a freshwater lake, into which terrestrial plants were drifted. This condition ceased, apparently in a gradual manner, and the surface of the land became covered by sea, in a calcareous deposit (the *calcaire grossier* as it is termed,) from which thousands of marine animals were entombed, few being analogous to species now existing. By the organic remains found in a gypsum rock above this last, we learn that numerous mammiferous creatures, belonging to genera which have ceased to live upon our planet, inhabited some neighbouring dry land. With these are found other mammiferous remains, belonging to existing genera ; but the species of all are alike extinct. We must refer to the works of the illustrious Cuvier for details respecting the species of Palæotherium, Anoplotherium, Chæroptamus, Canis, Didelphis, &c., detected in these beds. In all probability the climate in which they lived was then warm, judging at least from the palms and crocodiles which appear to have existed at the same period. Another change took place ; sea covered the area, and was again removed, subsequent freshwater deposits affording evidence of the presence of dry land. Somewhat similar events took place in the London and Isle of Wight basins at the same periods. The lowest part of the supracretaceous series in these places does not present the same freshwater character, for the freshwater remains are subordinate to the marine. We could scarcely expect it to be otherwise, for freshwater deposits must of necessity be more or less limited. During the time that calcareous matter was thrown down in the Parisian basin, as it is termed, mud was deposited in those of the Isle of Wight and London, mixed with sands. The organic remains still seem to point to a warm climate, at least if crocodiles be considered, from analogy, always to have required a temperature at least equal to that of Egypt. The vegetable remains have never been well examined, but it is supposed that among the abundance of fruits and seeds discovered at Sheppey, the greater part required a warm climate for their production. Freshwater remains are found above these marine deposits in Hampshire and in the Isle of Wight, and we have evidence in them that the extinct genera Anoplotherium and Palæotherium extended to that part of the European area, and that a species of ruminant allied to the genus Moschus inhabited the dry land with them. How far these French and English deposits, and certain others detected in Belgium, and also referred to the lowest part of the supracretaceous rocks of Europe, may have been continuous, it would be difficult to say; that they may have once been so is very probable, particularly the London and Belgian deposits, which seem to be only apparently separated by the intervening sea, or more modern rocks. That various disrupting causes have altered the condition of this part of Europe since these deposits were effected is quite clear, and therefore we probably find only a portion of those which once existed.

We are necessarily not to consider this part of Europe the only portion of the earth's surface where deposits of matter were produced at this period. It by no means follows, however, that they should be all characterized by similar organic remains, or by the presence of a certain per centage of particular molluscs; we should rather infer that they would not, unless we suppose equal conditions obtaining in every situation, where deposits then took place, over the whole superficies of our planet. The researches of Mr. Lea in North America have brought to light a series of fossils, principally shells, entombed in a deposit of Alabama, which the author suspects may be equivalent to the lower supracretaceous rocks of Europe, though it has not yet been found to contain any shell analogous to existing species. This author states* that out of nearly two hundred and fifty species taken from this rock, and examined by him, he does not consider one to be strictly analogous to those of the lower supracretaceous deposits of Europe. It is, however, remarkable, that this mass of shells does not resemble the organic contents of any other European rock, and therefore the inference rests upon the prevalence of certain genera supposed to have first appeared on the surface of the earth at this time. Here, therefore, we have a rock which may be equivalent to some group of the fossiliferous deposits of Europe comprised within a certain range, full of fossil-shells, yet they afford us little help beyond the supposition that particular genera may characterize a given portion of the fossiliferous rocks generally. The researches of Mr. Lea are nevertheless highly valuable, and, when added to those of other active American geologists, must eventually lead, in conjunction with those of European geologists, to more exact know-

* Contributions to Geology: Philadelphia 1833.

ledge respecting the value of organic remains in determining the relative antiquity of rocks, than can be accomplished by the labours of those, however distinguished and acute they may be, whose researches are confined to that part of the earth's surface now occupied by Europe.

We have not space to enter into a detailed account of the various supracretaceous rocks further than to observe that classifying them in the order of a certain per centage of existing molluscs in each, the medial portion is supposed to occur, according to Mr. Lyell, in the districts of Bordeaux, Dax, Touraine, Turin. Baden, Vienna, Moravia, Hungary, Cracovia, Volhynia, Podolia, Transylvania, Angers and Ronca, while the superior portion is found in Italy, Sicily, the Morea, and includes the English crag, a remarkable deposit occurring on our eastern coast. All the evidence tends to show that as these various accumulations of matter were produced, dry land became gradually more extensive throughout Europe, numerous spaces occupied by seas, lakes, or estuaries having been converted into the abodes of terrestrial creatures, either by being filled by deposited solid matter, or by being raised above the general level of the ocean. There is also abundant evidence to show, viewing the subject generally, that this change was accompanied by a gradual alteration in animal and vegetable life, so that it finally became such as we now find it in this quarter of the world. The supracretaceous group apparently passes so insensibly into the present order of things, still viewing the subject on the large scale, that probably no line of demarcation will ever be drawn between them, particularly when we regard the whole superficies of the world, and not a particular portion of it. Still, even during the time that the upper portion of the series was deposited, great changes must have been produced in animal life over a considerable part of the earth's surface.

Among the various mammiferous animals found fossil in the supracretaceous series of Europe, it is only among the more recent deposits that any question can arise as to whether they are or are not analogous to existing species. And even then there are few remains which can lead to a doubt of the whole mass of mammiferous life entombed being specifically distinct from that now living. Not only have species disappeared at this epoch, but whole genera have been swept away from among animals existing on the surface of our planet. The Palæotherium was a genus evidently somewhat common when the lower supracretaceous rocks were deposited, since Cuvier enumerates seven species as found in the gypsum beds near Paris. This genus was, however, by no means confined to that vicinity, for its remains have been detected in the Isle of Wight, various parts of France, in Germany, and Switzerland. Assuming that the Touraine *fahluns* have been correctly referred to the medial portion of the series under consideration, the Palæotherium then lived in the same districts with the Mastodon, Rhinoceros, Hippopotamus, Equus, and Cervus, animals which have not yet been detected in the lower supracretaceous beds. We therefore might conclude that the Palæotherium was a genus created before these latter creatures, and that it continued to exist after they were called into life. Similar mixtures of organic remains have been noticed by

Mr. Murchison and Count Munster at Georges Gemund, (where the exuviæ of the Bos and Ursus were also detected, the latter being the cavern bear,) and by M. Meyer at Friedrichs Gemünd and Eppelsheim. Whether these deposits may or may not be contemporaneous with the Touraine fahluns it may be difficult to say, but the facts are highly interesting, as they show that the Palæotherium lived at the same time with the animals above enumerated in other places than Touraine, and therefore that the mixture of their remains in the latter district was not accidental. Among the great abundance of bones found in different superior portions of the supracretaceous deposits, where they may be said to pass into those now taking place, the remains of the Palæotherium have not been found; we may therefore conclude that the genus ceased to live on the surface of the earth before the appearance of man.

The history of the Anoplotherium seems to be much the same as that of the Palæotherium: it disappeared like the former before the deposit of the supracretaceous rocks terminated, having first lived in the same districts with Mastodons, Rhinoceroses, &c. Other genera, found in the lower part of the series, such as the Anthracotherium, appear also to have perished in the same manner. All the genera, however, entombed when the lower part of the supracretaceous series was forming, have not been swept away from among living creatures; for we find Canis, Sciurus, Didelphis, and others enumerated by Cuvier as found with the Palæotherium and the Anoplotherium in the gypseous beds of Paris.

That huge creature the Mastodon, whose remains

are scattered over various parts of the world, appears to have been a genus created about the period that a medial portion of the supracretaceous rocks was deposited. This remarkable genus has ceased also to exist, though it probably continued to do so until a comparatively recent geological period; for the exact age of the beds in which the exuviæ of the *Mastodon maximus* is detected in North America cannot be considered as well determined. Indeed, if it be certain that a species of reed still growing in Virginia was detected in a kind of sack (supposed to be the stomach of one of these creatures,) among the bones of a Mastodon discovered at Withe, in the same country, it may have existed even up to the appearance of man on the surface of this planet. *A priori,* we do not readily perceive why, when Elephants and Mastodons evidently existed together in many parts of the world, the former genus should still exist, while the latter should have perished. We should be disposed to consider, from analogy, that a change of climate alone would have been as fatal to Elephants as to Mastodons, particularly when those of hot countries, such as that of India, are concerned, and where we have no reason to suppose the temperature to have then been much greater than at present. Yet Mastodons are found in a fossil state in India and equinoctial America as well as in the colder regions of Europe and North America.

Other huge creatures belonging to extinct genera, which, as far as the evidence extends, were called into existence after the Palæotherium, have also disappeared from among existing animals. As yet, however, we do not know that they have been extensively distributed, like the Mastodons; for the Megatherium has only been detected in South America, and the Deinotherium on the banks of the Rhine.

While some genera have thus been successively created and destroyed, others, apparently also created in succession, have been continued to the present day. We have the Elephant, Rhinoceros, Hippopotamus, and others entombed in various supracretaceous deposits, still existing as genera among living animals. The species, indeed, are distinct from the fossil, and they are not discovered in many regions where numbers, judging from the amount of their remains, must once have roamed in the situations best suited to them. There has been a change in their distribution over the surface of the globe, but the generic form has been retained. With regard to many, the surface occupied by them is more limited, and their places have been supplied by the creation of animals better suited to the present condition of the earth.

We cannot close this slight sketch of the fossiliferous deposits without adverting to those extensive tracts of slightly inclined planes bounded by sea-coasts, which occur so commonly in various parts of the world, and which are found, wherever examined with attention, to contain numerous molluscs analogous to those existing in the seas which skirt them. The per-centage of such molluscs may perhaps vary, but there is something very remarkable in these extensive tracts of nearly level country which slope gently into the sea in so many parts of the world, and which are generally bounded by considerable areas of soundings. They seem, indeed, mere continuations of the latter, which have been relatively raised at comparatively recent geological epochs above the waters of the ocean. If the ocean-level were raised, or the continents depressed, 600 or 800 feet, a large portion of dry land, formed of these plains, would be submerged. These facts lead us to inquire whether any condition common to the earth's surface, or crust, at a particular period has caused such similar general appearances, or whether they may be due to a series of effects produced at different times which have finally led to similar results. In Europe, taking into account every probable loss from the action of seas on coasts, there seems good evidence of a gradual increase in dry land generally, during the supracretaceous period, from changes of level, and by the deposit of solid matter in various seas and lakes which have ceased to exist. As far as our knowledge extends, similar effects are observable in other continents. Now if this be true generally, there would be a gradual increase of dry land over the surface of the earth at the same epoch; and since we may infer an increase in the relative amount of terrestrial creatures generally at the same period, the one would harmonize with the other, and we should obtain an amount of terrestrial life called into existence in proportion as dry land increased and was fitted for its support.

CHAPTER XVIII.

During those millions of years that we should conceive necessary for the production of the various deposits of which we have given a brief sketch in the three preceding chapters, other rocks in a state of fusion have been ejected from beneath, upheaving, overlapping, or forcing their way among, those beds which from analogy we refer to an aqueous origin. It was at one time supposed that the particular mineral compound named granite was peculiar to the lowest portions of the rocks composing the crust of the globe, that, in fact, it constituted the fundamental rock upon which all others had been formed, and was not discovered higher in the series. This opinion, like many others connected with our subject, has given way before facts, for we find granitic rocks in situations where they must have been ejected subsequently to the period during which the cretaceous group was deposited, as also in other places, into which they must have been thrust at intermediate periods down to the oldest rocks inclusive. The mere fact of finding granitic rocks high up in the fossiliferous series is, however, no proof that the conditions necessary for their production have been alike favourable at all times, and up to the supracretaceous epoch inclusive. We should study the abundance or rarity of the rocks in question at different geological periods, and thence infer the more or less favourable conditions for their production at such epochs according to the general nature of the evidence we should thus obtain. When we proceed in this manner, we soon arrive at the conclusion, that during those periods when the inferior stratified or non-fossiliferous rocks were formed, the amount of granite produced must have been greatest; and when we further proceed to examine the differences which may exist between such rocks and the various granites associated with them, we find that in many there is no chemical difference whatever, and in many others one that is extremely trifling. As a mass, the various granites and the inferior stratified rocks bear a close chemical resemblance to each other; as a mass, they are silicates, and are usually associated in a manner which would lead us to suppose that, at the geological epoch when the whole was formed, there were conditions on the surface of our planet highly favourable to the production of a few remarkable mineral substances.

In whatever manner we suppose the inferior stratified rocks to have been formed, time was essential for their production; and as there are many reasons for concluding that a great lapse of time was necessary for the purpose, we may fairly infer that numerous masses of granite were included among them at different relative periods. Hence we should by no means expect to discover constant evidences of force when granite is associated with such rocks. Even considering all the granites so associated to have once been in a state of fusion, it is clear that they may have formed thick beds as overlapping masses, which have subsequently been covered over by mica slates or other rocks of that kind; or they may have forced their way through lines of stratification, being also those of least resistances. In either case we might obtain tabular masses of granite quietly associated with the inferior stratified rocks without marks of violence. That there have been violent intrusions of granite among these rocks is, however, also certain, since veins of the former frequently cut the latter in all directions, and there are sometimes evidences of movement, and of the application of force at the junctions of the rocks.

We have seen that fluidity was necessary to the figure of the earth, and we have inferred that this fluidity was igneous; hence, therefore, we should anticipate that, during those periods when the crust of the globe first became solid, there would not only be a large amount of crystalline rocks produced, but that there would also be a somewhat abundant intermixture of igneous with aqueous products, when water could exist as a liquid on the surface of the globe, and assist in producing mineral compounds constituting rocks. There is another important circumstance to be observed, which is, that while passages of gneiss into granite can be observed among these rocks somewhat on the large scale, those that have been supposed to take place from granite into more modern rocks, among which the former has been intruded, are comparatively insignificant, if even they can be said really to exist. Now if it be supposed that gneiss has been to a certain extent an aqueous product, a supposition which by no means follows from the arrangement of its component minerals, we can readily conceive that, in this heated state of the whole superficies of the globe, the igneous and aqueous products would have a tendency to be more blended than when fused rocks were ejected among cold rocks or into a cold ocean.

There is so much solid matter of a similar kind, whether termed gneiss, granite, or by whatever name it may be known, in the lowest portions of those rocks composing the visible crust of the earth; and so many deposits are evidently derived from abraded portions of it, that we are led to conclude the granitic was the form which matter assumed when it first became solid on the surface of the earth. It, however, by no means follows that granite necessarily extends to great depths beneath the superficies of our planet; on the contrary, assuming the igneous fluidity of the globe, we should expect that the substances most readily driven to the surface would be those first crystallized, affording at the same time materials, by abrasion and solution, for deposits by water, when that substance became an agent in the formation of rocks. Hence we might infer, that when by a sufficient radiation of heat a considerable amount of the earth's crust became solid, matter somewhat different from the granitic might be ejected through any fissures which existed, and possibly that such matter might possess, under equal circumstances, somewhat greater density.

Supposing this to be the case, we should consider that granitic matter would be thrown up at different periods; but we should expect that it would eventually become mixed with other matter, originally and as a mass situated beneath it, and that it would ultimately be an exceedingly rare product, except under circumstances which should produce the fusion of previously consolidated granite, gneiss, or analogous rocks. Now this would happen from variations in the depth of the line of temperature beneath the earth's surface proper to cause such fusion, which variations would be produced, in accordance with the views of Mr. Babbage, by those in the surface itself causing changes in its power of radiating heat. We may obviously obtain the fusion and ejection of previously consolidated granitic matter by mere local causes, but the simplicity of a principle which, when coupled with central heat, applies to the whole surface of our planet, not confined to minor portions of it, and which would be productive of so many effects, is so striking, that we are led to consider it a power capable of causing numerous and important geological effects. Let us, for instance, suppose that the line of elevated temperature capable of fusing granitic matter, passed beneath a mass of gneiss; that through a fissure, produced by fracture from any of the causes noticed in other parts of this volume, a sufficient quantity of greenstone or any analogous rock was ejected, spreading over the surface above; the line of given temperature, supposing heat to exist in the interior of the earth, would rise, particularly if the surface be changed from sea to dry land, and the gneiss be fused up to the new line of given temperature; there would, therefore, be a mass of granitic matter in fusion ready to be thrown upwards at the first period when conditions should be favourable, such as a continuance of those causes which produced the original fissure, and which subsequently again broke the crust of the earth at that point.

When we regard igneous rocks on the large scale, we are struck with the fact, that while they all appear a mixture of certain silicates, silica itself is more abundant in the granitic than in the trappean rocks, (as greenstones and the like are commonly termed for convenience,) at least calculations respecting their chemical composition would lead to this view*. Now if we infer, and it appears to be a fair inference, that granite, considered as an igneous rock, was more abundantly produced at the earliest periods than at any other, we arrive at the conclusion that silica was not only very abundant among the stratified rocks of the period, no matter how produced, but also among the massive rocks of the same period, and hence that, during the consolidation of the crust of the earth, silicium became one of the earliest oxidized substances, and then produced, by combination with other substances, the greater amount of silicates observable in the solid crust of the globe. How silica, after a certain thickness of rocky matter, may diminish in quantity in the direction of the earth's centre, is another question; but if we suppose the matter of trappean rocks, viewed in the mass, to have been situated beneath the granitic, and to have been expelled in proportion as the latter became solidified, we should infer that such might be the case.

While, however, the silica diminishes in trappean rocks, viewed generally, the proportion of magnesia and lime, particularly the latter, increases;—another remarkable analogy between the mass of aqueous and igneous rocks of two very different geological periods, for lime is certainly more abundant among the more modern than among the more ancient rocks, viewed

* Geological Manual, p. 448.

as masses of matter. We naturally inquire if the lime thus distributed by different means over the surface of the earth, more at one time than at another, may not be due to a communication between such surface and matter situated beneath it, greater at one period than at another; and whether the consolidation and subsequent fracture of the substances which previously prevented this communication, at least to the extent required, may not have been one of the principal causes of such phænomena.

The presence of hornblende, a mineral in which lime amounts to about 13·5 per cent., and magnesia to about 18·5 per cent., is the principal cause of this difference in the composition of igneous rocks. It is a much more common substance among the more modern than among the ancient rocks of this class; not that it is absent from the latter, for granites, apparently of considerable geological antiquity, contain it, and hornblende rocks are found sometimes in considerable abundance associated with gneiss, mica-slates, and others of the inferior stratified rocks. There can, however, be little doubt, that, viewed in the mass, the hornblende and augitic rocks prevailed at the later geological epochs.

Nothing is more common among igneous rocks than to observe a change of mineralogical structure in the same mass, and at short distances. We may find greenstone, porphyry, and sienite in the same hill, each passing into the other so gradually that it is difficult to say what name should be applied in particular to the mass, and hence geologists find it convenient to affix some general name to these rocks, such as trappean, to avoid the extreme difficulty, almost amounting to impossibility, of presenting a clear idea of their variable structure in a few words. Sometimes this difference in mineralogical structure is more apparent than real, that is, the chemical composition of two igneous rocks may be the same, and their mineralogical appearance be different; the apparent difference arising from the conditions under which they have respectively been placed, the one perhaps taking a porphyritic form, the other one more confusedly crystalline. The more gradually a mass of fused rock cooled, the more perfect we should expect to find the crystalline arrangements of its parts. Now this is what we do find in experiments, imperfect as they necessarily are, intended to illustrate this point. When, therefore, veins, such as those of granite, greenstone, or any other igneous rock, diverge from masses of the same substances into contiguous rocks, which have evidently preexisted, and have been fractured prior to the intrusion of the igneous matter of the veins, we most frequently find that the mineralogical arrangement of the constituent substances differs in the veins, and in the mass whence they diverge. It is particularly interesting to trace granite veins, as may under favourable circumstances be done, from granitic masses with marked characters, such as those of Devon and Cornwall, until the crystalline character be finally lost; so that, unless we could so trace them, we might doubt the continuity of a rock, having perhaps the appearance of a compact felspar, with granite, in which the crystallization was so advanced that large crystals of felspar, or albite, were isolated in the granite by thousands within a few cubic fathoms of rock.

It hence follows that, all other things being the same, the worse the conductors of heat by which a rock in fusion is surrounded, the more crystalline will be the structure of its parts, and therefore, assuming central heat to exist, and that rocks generally are equally bad conductors of heat, the deeper beneath the surface of the earth's solid crust a rock in fusion may be ejected among others, the more crystalline would it become when finally consolidated. It must not hence be inferred, as has been done, that herein consists the only difference between the various kinds of igneous rocks, and that greenstone would have been granite if it had taken a longer time to cool. Granites no doubt vary in their chemical composition, and so do greenstones, yet they always so differ from each other as masses of matter that the one can never become the other from mere differences in cooling. This difference will be apparent if we take, by way of illustration, a granite composed of two fifths of quartz, two fifths of felspar, and one fifth of mica, which may be considered an average common kind of granite, and a greenstone, composed of equal parts of felspar and hornblende, an average kind of greenstone, and observe the differences between their calculated chemical compositions *.

	Granite.	Greenstone.	Difference.
Silica	74·84	54·86	19·98
Alumina	12·80	15·56	2·76
Potash	7·48	6·83	0·65
Magnesia	0·99	9·39	8·40
Lime	0·37	7·29	6·92
Ox. of Iron	1·93	4·03	2·10
Ox. of Manganese	0·12	0·11	0·01
Fluoric Acid	0·21	0·75	0·54

* Geological Manual, pp. 448—450.

As inferior rocks, whether stratified or unstratified, may become fused and thrown up from local causes, such as the percolation of water to the metallic bases of certain earths and alkalies through cracks in a part of the world's solid crust, long unbroken in that particular part, or from the more general cause of alteration in lines of given temperatures from changes produced in the radiating powers of the surface above them, we are not certain whether trachytic rocks, which so frequently appear chemically to approach granites, are derived from matter previously consolidated in the form of rocks, or whether they constitute a portion of the original granitic matter that has not been consolidated prior to its ejection from those volcanic vents whence trachyte has issued. In some volcanic districts trachytic rocks are, as a mass, found to occupy an inferior position to the basaltic rocks, also taken as a mass. Now we may obtain this result in two ways; inferior granitic rocks or gneiss may have been fused and ejected before the basaltic matter was thrown out in abundance; or granitic matter may have floated to a certain extent, from its inferior specific, gravity on the basaltic matter, and therefore, being nearest the orifice when formed, was the first to be thrown up through it. In some districts trachyte seems little else than granite which has been again fused, and having been exposed to different conditions, no longer presents the appearance of granite. Indeed, specimens from Auvergne exist which are partly granite partly trachyte, the one structure shading off into the other.

When granitic, greenstone, basaltic, or other masses of rock of a similar kind exist on a given portion of the earth's surface, and either from local or more general causes gradually become heated beneath, more at one point than at others near it, the lines of temperature capable of producing a viscous or tenacious condition of the rock, rising continually though slowly upwards, such rocks would bulge outwards or be elevated in dome-formed masses; first from the increased volume caused by the increase of heat more at one point than another, and secondly, from the action of any elastic matter endeavouring to escape. As trachytic and other dome-shaped masses of a similar kind are found in volcanic countries, the action of elastic matter endeavouring to escape, and assisting or perhaps wholly causing their production, would be in accordance with volcanic phænomena generally.

Now assuming that a mass of solid matter is rendered yielding by a gradual increase of heat, and that elastic matter is compressed beneath it, when the force of the latter can lift the former, the amount of the elevation produced will depend upon the power of the elastic matter on the one hand, and the yielding state of the rocky matter on the other. If the power of the elastic matter be considerable, and the rock yields to it, much would depend upon the manner in which the latter did thus yield: if it were very viscous or tenacious, it might be blown upwards into a dome-formed mass, while if it were to a certain extent brittle, it would be fractured, and the elastic matter escape. As the elastic matter expanded beneath such a dome, it would necessarily become less capable of lifting the superincumbent rock than it was in its greater compressed state before the dome was produced, and the substance of the rock itself, composing the dome, being removed from the greater heat beneath, would tend to become less yielding; these two causes, therefore, would combine to arrest the elevation of the dome beyond the point where the various powers should balance each other, and we should obtain a result not very unlike those trachytic and basaltic domes which seem to have been forced up much in the condition we now find them, proper allowances being made for the subsequent action of meteoric influences upon them.

Very different degrees of heat seem necessary for the fusion of different igneous rocks, the most refractory appearing to be the serpentines, and the most fusible the basalts and various lavas. In a general sketch like the present, the serpentines and rocks of that family may be neglected, even though they occur in considerable masses in some countries in a manner to leave little doubt of their igneous origin in such situations; for, viewed generally, their collective volume is very inferior to the mass of granitic and trappean rocks. Now the two latter differ considerably in fusibility, the granitic rocks being by far the most refractory, so that if we consider the greater mass of granitic rocks to have been produced at those early periods when the surface of our planet first became solid, igneous rocks ejected at comparatively recent periods are compounded of more fusible materials than those forced upwards at more ancient epochs. The presence of a far larger quantity of silicate of lime in the trappean than in the granitic rocks, seems the principal cause of this difference. It would, however, follow from this difference, that at a given depth beneath the surface of the earth's crust, a certain amount of heat would be sufficient to keep trappean matter in fusion which would be insufficient to produce the same effect on granitic matter, and hence that granitic matter could not be ejected from such given depths, though when the heat was greater, and consequently such matter fusible at the same depths, it might have been the first thrown out, being of less specific gravity and probably nearer the surface. We should expect occasionally to discover mixtures of trappean and granitic matter, even constituting one mass, and these are not unfrequently found, the one passing insensibly into the other, the mica, a most refractory substance in itself, disappearing and being replaced by hornblende, a readily fusible mineral. The age of many of these masses requires to be better determined than it has been, before we can see how far it may affect the hypothesis, that the great mass of granitic matter has been ejected at the more ancient, and the mass of trappean matter at the more modern geological periods.

When we regard the fossiliferous rocks, we appear to find evidence that, from the time they were first produced, dry land and shallow seas existed. Hence we might conclude that during such periods subaërial eruptions of igneous matter then also took place, and that elastic vapours being also discharged, ashes, cinders and dust would be thrown up and transported around the igneous vents. From the various surface-changes which have taken place on our planet, the evidences of any such eruptions are necessarily difficult to trace, and would become less, viewing the subject generally, in proportion to the antiquity of the series of rocks among which we seek them. There are, however, certain situations where we find igneous rocks, more particularly trappean, that appear contemporaneous with those among which they occur, that is, they appear to have covered an inferior bed, and to have been subsequently covered by the tranquil deposit of transported or other matter above them; precisely as a bed of lava may flow over a sandy bottom, and afterwards be covered up by a deposit of sand or mud. Trappean rocks are, as is well known, much associated with the lower parts of the grauwacke series in different parts of Europe, sometimes in a manner which leaves little doubt that a portion of them has not been intruded among the strata subsequently to their consolidation, but has been formed in the manner above noticed; while others, again, have clearly forced a passage through the grauwacke, sometimes even cutting the former. It is frequently interesting to observe that such beds of greenstone or porphyry, though thick in parts, fine off among the grauwacke beds by taking the character of an arenaceous deposit of comminuted trappean matter; as if such portions constituted a deposit of trappean ashes thrown out at the same time that the trappean rock itself was produced.

We have had occasion to observe considerable accumulations of such comminuted trappean matter among the greenstones and porphyries of the older grauwacke of Devon and Cornwall, not only in lines continuous through the general stratification of the grauwacke of the district, but also associated with them in a manner which might lead us to suspect that there had been eruptions at those spots, either into the atmosphere or beneath shallow water, permitting ashes to be thrown up and dispersed. One spot, Brent Tor, north from Tavistock, is particularly remarkable, as a conglomerate there occurs some of the constituent parts of which have every appearance of volcanic cinders, an infiltration of siliceous and other matter having, during the lapse of ages, filled the vesicles of the cinders. The dip of the strata in that part of the country not being so considerable as it usually is in the district, these beds can be traced for some distance, and are observed to fine off in a remarkable manner into the arenaceous trappean rocks above noticed. The country for several square miles around is either composed of arenaceous grauwacke, greenstones, porphyries, or the arenaceous trappean substance apparently so like an ancient deposit of volcanic ashes. As from accidental circumstances, which have no relation to its general geological structure, Brent Tor has the appearance of a volcanic point, which might be considered comparatively modern, it may be necessary to state that the various rocks noticed can easily be traced into the grauwacke of the district, with which they are evidently associated.

We have entered into the foregoing detail merely to show, that among the oldest fossiliferous rocks there are facts which would lead us to suppose, that there had been ejections of igneous matter into the atmosphere or beneath shallow water, and consequently that we might expect to discover similar facts among the other fossiliferous rocks, under favourable circumstances and in different parts of the world.

The various fossiliferous deposits have been interrupted not only by these and other larger ejections of igneous matter, but also by those comparatively great dislocations of the earth's crust which have produced ranges of mountains, and tilted and contorted strata over large areas, thus altering the condition of our planet's surface over such areas, and producing a variety of consequences, different from those which resulted from the prior condition of the same portions of surface. M. Elie de Beaumont considers that he can recognise at least twelve principal periods of dislocation in Europe, which have been succeeded by as many changes in the sedimentary deposits of the same area. Without at all entering into the evidence adduced on this head, for which we must refer the reader elsewhere*, we should anticipate that if mountain-ranges have originated from any causes analogous to those noticed in a former part of this volume, they would be eventually found to be more numerous, interspersed, as it were, among a great number of minor dislocations, resulting from a less intense exertion of the same forces, such as that observable in common earthquakes. All the evidence, indeed, respecting the latter, shows us that the force by which they are produced varies in intensity; and as they are felt in all parts of the world, as well in those situated at distances from volcanic vents, as in those in their immediate vicinity, we may fairly infer that their cause is situated beneath the surface of the earth generally, though from local circumstances they

* Recherches sur les Révolutions de la Surface du Globe; Annales des Sci. Nat. 1829 et 1830. The reader will also find an account by M. Elie de Beaumont of his present views on this subject, in the French Translation of the Geological Manual; p. 616. Paris 1833.

may be more intensely felt in one part of it than in another at different geological epochs, being generally most intense where, for the time, volcanic vents exist. From the theory of central heat, and that of water percolating to certain metallic bases of the earths and alkalies, which may be termed the volcanic theory, we obtain two causes capable of producing disruptions of the earth's solid surface. Such minor movements as those termed earthquakes may originate from either, while the more general and vast dislocations, productive of long lines of mountains, would appear referrible to the gradual refrigeration of the globe. We should, indeed, expect that volcanic action would considerably modify the effects produced by these greater disruptions, since conditions would then be favourable to it, volcanos appearing on the great lines of fracture; but it does not seem, taken by itself, equal to the production of those long lines of elevation constituting such mountains as the Himalaya, the Alps, and others.

We should expect to discover the effects produced by comparatively considerable disruptions of strata, better exhibited among the more recent states of the earth's surface than among such traces of the more ancient as those we can observe, since the former have been necessarily exposed to fewer causes of obliteration than the latter, such as the accumulation of strata upon them, and others. Independently of the great contortions and fractures of strata so frequently observable, we find masses of rock of considerable volume and weight detached from particular places and transported long distances. These masses of rock are generally known by the name of *erratic blocks*. In Europe we have at least two accumulations of them produced, judging from their geological position, at comparatively recent periods. One set of erratic blocks has been scattered from the central Alps outwards, on each side of the chain; the other has proceeded from a northern direction southward. How far the events which have produced both accumulations of these blocks may have been separated by time from each other, we know not; but we are certain that the geological epochs of both must have been very recent, since they both rest on rocks of little comparative antiquity. That the principal valleys of the Alps existed nearly as we now see them appears also certain, since the blocks derived from the central chain have been borne down through them, leaving abundant evidences of their transport on their sides, while the principal accumulations are in front of such valleys. M. Élie de Beaumont has suggested a very simple explanation of the transport of the Alpine blocks. He supposes that during the last elevating movement which took place in the Alps, the heat evolved from the necessary fissures suddenly melted the snows which previously existed on these mountains, and by these means a large body of water was produced, which swept the blocks through the valleys into those situations where we now find them. If to this we add the probability that the glaciers (supposing the necessary height of the mountains,) would be covered as now by multitudes of blocks, we certainly obtain, by means of this ready mode of transport, and the sudden action of a large body of water rushing down the valleys, a more plausible explanation of the phænomena observed, than by the supposition of a huge mass of water overwhelming the whole land, which does not afford us a good explanation of the nearly equal transport outwards, down so many principal valleys, of solid matter derived from the central parts of the chain. The only question likely to arise under this hypothesis is, whether the sudden melting of the Alpine snows, even as they now exist, would produce a volume of water sufficient to cause the effects observed. In some favourable situations, such, for instance, as the Monte San Primo, rising above the Lake of Como, there is evidence of a tumultuous deposit of these blocks with smaller detritus, the whole being mixed pell-mell together; in others, again, we find only large blocks, as on the Jura, perched in lines high up the sides of mountains. M. Élie de Beaumont has shown that in the valley of the Durance the blocks decrease in volume and become less angular as they recede from the mountains behind Gap, until the transported matter diminishes to the pebbles constituting the wide extent of country known as the Crau. Similar facts are also observable down other valleys. Perhaps in the *loess* of the Rhine we may trace the remains of still finer detritus, which has accumulated to the depth of 200 or 300 feet above the valley, and bears evident marks of sudden transport. This supposition is rendered more probable by the abundance of Alpine pebbles discovered resting on various rocks, where the loess ceases in the higher parts of the valley of the Rhine, and which have apparently been accumulated by a sudden rush of water down the valley.

The other great accumulation of erratic blocks seems due to some more general cause, since not only are the blocks scattered in great abundance over Northern

Europe, in a manner to show their northern origin, but those which occur in the northern parts of America, apparently in equal abundance, also point to a similar origin *. We hence infer that some cause, situated in the polar regions, has so acted as to produce this dispersion of solid matter over a certain portion of the earth's surface. We know of no agent capable of causing the effect required, but moving water. We therefore further infer that some cause has produced an action of water in the polar regions which has scattered blocks of rock outwards from a somewhat central situation. Now, if mountains can be produced in the manner noticed in a former part of this volume, there is no reason, that we are aware of, why a sudden elevation of land should not be produced beneath the polar seas as in any other part of the earth's surface, and if such elevation were so produced, the necessary consequences would be a wave or waves proportioned to the disturbing force. Such waves would necessarily tend to float the northern glaciers with their usual burden of blocks of rock, lifting them to the southward; but their principal action would be felt where they reached the coasts, and the waves from being little more than great undulations of water, became huge breakers, acquiring a violent forward motion, and a consequent abrading and transporting

* Professor Hitchcock has presented us with a very detailed and valuable account of the erratic blocks in the state of Massachusetts, (Report on the Geology of that State, p. 141 to p. 171. Amherst 1833.) where, with detritus of minor volume, named *diluvium*, they occur in considerable abundance. The proofs he affords of the general northern origin of the whole transported mass are highly satisfactory, coinciding with the evidence afforded by the researches of Dr. Bigsby, Messrs. Lapham, Jackson, Alger, and others, in various parts of North America.

power, which can be well appreciated when it is recollected that those relatively minute waves which break on coasts during gales of wind have the power to destroy large solid pieces of compact masonry, and to hurl the disjointed fragments before them. The effects observed would correspond with this hypothesis, for all the blocks have not come from great distances: they have been detached from various points. Many erratic blocks in England can be traced northwards to their parent rocks in the British Islands, and Prof. Hitchcock has shown that the like can be done in the United States *. The great mass of blocks scattered in such abundance over parts of Germany, Sweden, Poland and Russia, are evidently derived from rocks known to exist to the northward of them, their mineralogical and other characters being sufficiently marked. Many of the effects which would be produced by the breaking of such waves will be obvious; among the rest we should expect considerable local accumulations of rocks derived from different quarters. This seems to be well shown in the diluvium of Holderness, on the coast of Yorkshire, where, according to Mr. Phillips, a clay envelops various fragments of rocks, apparently derived from Norway, the highlands of Scotland, the mountains of Cumberland, and from Yorkshire itself, while no small portion seems to have been transported from the seacoast of Durham and the vicinity of Whitby; the fragments being rounded in proportion to the distance whence they were derived.

As a whole, we may consider the facts connected

* Report on the Geology, &c., of Massachusetts, *Art.* Diluvium. Amherst 1833.

with erratic blocks highly important, whatever theory we may adopt, and which shall at the same time afford a fair explanation of the observed phænomena, since they teach us, in conjunction with those large contortions of strata, and the great dislocations of rocks so often observable, that while we duly appreciate the continued and more tranquil geological effects we daily witness, we should not neglect those other effects, which, though apparently great when measured by the ideas commonly entertained on such subjects, or by the effects produced by minor intensities of the same forces, are yet comparatively insignificant when considered with reference to the spheroid on the surface of which they occur.

Recent astronomical researches have rendered it extremely probable, indeed almost certain, that the planets move through a resisting medium. Assuming this to be true, it follows that all such bodies with their attendant satellites must eventually fall into the sun ;- and consequently that there is no real stability in the solar system, it being one of constant though quiet and long-continued change. Myriads of those revolutions of the earth round the sun by which man measures long periods of time may elapse before our globe ceases to move as a planet ; but if this resisting medium does exist, and the general course of events is not interrupted by an extraordinary cause, it must ultimately constitute a portion of the sun, itself probably moving among other bodies of equal and larger volume, and destined, for aught we know, to merge with others like itself into some greater mass of matter.

We cannot contemplate this termination to the existence of the earth without perceiving that a slow but certain change in the condition of its surface must of necessity follow, even supposing that no cause of change exists in the earth itself. All animal and vegetable life now existing on the globe is most beautifully adjusted, as Mr. Whewell so forcibly expresses it *, to the conditions under which it is placed. It may be probable that a large portion of existing animal and vegetable life might accommodate itself to a certain amount of change; but there would evidently be a limit to this power of adaptation, and such life, suited to given conditions, must perish when those conditions no longer exist. It is obvious that, when the orbit of the Earth became no greater than that of Venus, great change of conditions would be effected on its surface, and that there would be still greater when its orbit became less, so that if it were covered with life as we should anticipate it would be, we should infer that such life must be accommodated to the new conditions under which it would be placed. And we can as readily conceive the existence of life so adapted, as that animals and plants have been created to live in the present hot and cold regions of the earth's surface respectively, and which would perish if removed from the conditions to which they have been adjusted.

The probability of future change from external causes leads us to look back on that which geological phænomena attest has already been produced on the surface of the earth. Whether we consider, with La Place, that our solar system is a condensation of nebulous matter, resembling those masses which astro-

* Bridgewater Treatise; On Astronomy and General Physics.

nomers have shown us to exist in various portions of the vast regions of space, the planets and their satellites having been separated from the principal part of the mass which now constitutes the sun; or that the sun, planets and satellites were originally created in those relative situations where we now observe them, existing knowledge leads us to infer that liquidity was necessary to the figure of the earth, and that great heat was necessary to such liquidity. As far as we can perceive, we should perfectly agree with Mr. Whewell in considering that the evidence of design would be quite as great, if gaseous matter had by condensation produced the solar system, as if it were created such as we now find it. There is, however, this difference between the hypotheses, the former presents us with a far more noble and splendid view of the great system of the universe than the latter, and is therefore one which, *à priori*, we should feel more disposed to adopt.

If the earth were one fluid, and its fluidity due to the action of heat, it follows, from the figure of the globe and the known laws of heat, that its surface would by radiation become solidified, and even comparatively cold, while intense heat still prevailed in the interior. We have direct evidence by ascending in our atmosphere that heat decreases upwards; and the calculations of Fourier and Svanberg, made upon different principles, afford a temperature of about $-58°$ Fahr. for the planetary spaces. A time must therefore arrive when the surface of the earth, exposed perhaps for millions of years to this temperature, would become comparatively cold, if even we suppose the heat contained in the mass of the globe itself to have once been sufficiently intense to cause a gaseous state of all the matter contained within it. Numerous chemical combinations would be gradually effected, and finally, the surface would be so circumstanced that it would depend upon the influence of the sun for a temperature beyond that it would otherwise possess, and consequently would be warmer in the equatorial than in the polar regions.

From our knowledge of the matter composing the visible portion of the earth, we should infer that a decrease generally in the heat of the globe would produce a diminution of its volume, and hence that the velocity of its movement would be increased. This alone would afford a cause of change affecting life on the surface of the earth; but when we couple it with a gradually diminished surface temperature arising from interior causes, and the consequent increased influence of the solar heat on the one hand, and the action of the low temperature of the planetary spaces on the other, we obtain a series of conditions which would seem to render it almost impossible, consistently with the constant adjustment of life to situation which we everywhere perceive around us, that any given set of animals or plants could continue to exist through all the changes which of necessity would be produced. Now the evidence derived from the organic remains entombed in the fossiliferous rocks, shows us that a great change has been effected in the animals and plants which have existed on the earth during the great lapse of time required for their deposit. Slight oscillations, as it were, may perhaps here and there be eventually discovered, when the geological structure of the earth's surface shall be more extensively explored, animals and plants having continued to exist in some situations longer than in others; but it does not appear in the least probable that the present evidence, indicative of a great system of change in animal and vegetable life, since such life existed on this planet, will be vitiated by new discoveries of this kind.

We are far from inferring that numerous general forms, particularly those of molluscs, discovered among the more ancient fossiliferous rocks, have not been continued to the present day; but, viewing the subject on the large scale, we consider that existing evidence points to a great system of change, generally as respects species, and frequently as regards genera, from the ancient state of animal and vegetable life to that now known to us. Neither do we suppose that minor changes may not have been somewhat suddenly produced; indeed the occurrence of extinct species of animals with their flesh preserved, incased in ice or frozen mud in the polar regions, and analogous to genera which now exist in warm countries, such as the elephant of the Lena and the rhinoceros of the Wiluji, (species moreover which must have existed, judging from their fossil remains, by thousands over various portions of the northern hemisphere,) appears to be sufficient evidence of this fact.

The evidences of change are, however, not confined to organic structure; the earlier rocks differ as a mass from those produced at later periods, also taken as a mass. The principal chemical difference between them consists in the extreme rarity of carbon in the former, and its comparative abundance in the latter, while the amount of lime is far greater in the more modern than among the more ancient strata. There are also other differences, but the most striking consists in the absence of organic remains in the one, while the other teems with those evidences of pre-existing life, often indeed to such an extent, that if such exuviæ were abstracted from the fossiliferous rocks generally, a great diminution of their collective volume would be effected.

Measuring time as man does in the minute manner suited to his wants and conveniences, a few thousand revolutions of the earth in its orbit appear to him to comprise a period so considerable, that he feels it difficult to conceive that great lapse of time which geology teaches us has been necessary to produce the present condition of the earth's surface; and which when we even take some great astronomical period as an unit, we still find it impossible to express. That man is a comparatively recent creature on the surface of the earth, all geological evidence tends to prove, since neither his remains nor his works have been detected except in the most recent deposits. From all analogy we may conclude that he was the first creature possessing any extended amount of intelligence which has existed on this planet. What new creation may supply his place, when, if the theory of a resisting medium in the planetary spaces be true, as it appears to be, conditions shall arise under which he cannot exist, it is of necessity impossible for us to form the most remote conception; but that he must, constructed as he now is, disappear from the surface of this planet, as other creatures have disappeared before him, should the external causes we have noticed be true and continue uninterrupted, even supposing no change to be produced by internal causes, seems certain. Consistently with that wisdom and design so observable in all the works of creation, we should infer that, so long as it were possible for life to exist upon it, this globe would no more revolve in space without some modification of life on its surface, than we can conceive the earth to be the only planet on which life exists. We may therefore conclude, that whatever changes our planet may suffer, either from external or internal causes, and the necessary conditions exist, life will be created to suit those conditions, even after man, and the terrestrial animals and plants contemporaneous with him, may have ceased to live on the surface of the earth.

APPENDIX.

Table of the Situations and Depths at which recent Genera of Marine and Estuary Shells have been observed. By W. J. Broderip, *Esq., F.R.S. V.P.G.S., &c.*

My friend Mr. De la Beche having requested me to form this table confining it to the situation, depth and bottom where marine and estuary shells have been observed, without regard to their geographical distribution, I am desirous of acknowledging the assistance which I have derived from Mr. G. B. Sowerby and Mr. Cuming. The information contributed by the last-mentioned gentleman is especially valuable, in as much as it arises from personal observation made during the voyage which has so much enriched this branch of Natural History. Still, notwithstanding these aids, I could have wished to have delayed the publication, till it had assumed a form which might have rendered it more complete and of more Geological value.

When the cross (+) occurs, it indicates that the depth is either unknown or that the information respecting it is deemed unsatisfactory; when 0 appears, it denotes that the genus, after which it is placed, is found, sometimes, at or near the surface of the water.

Annelids.

Serpula (including Vermilia and Galeolaria, *Lam.*). Generally littoral, attached to rocks, stones, shells, crustaceans, corals, and other marine bodies.

Spirorbis. On sea-weed, shells, &c.; and nearly in the same situations.

Sabella. Coasts, and on shells, &c.; generally in shallow water.

Terebella. Nearly the same situations.

Dentalium* Sometimes in deep water, frequently near the shore.

Cirrhipeds.

Pollicipes. Generally adhering to rocks in shoal water.

Pentelasmis. Attached to rocks, &c.; frequently found floating far at sea on drifted timber, net-corks, Ianthinæ, glass bottles, and also adhering to the bottoms of ships.

Scalpellum. Adhering to corallines, &c.; generally near shores.

Otion. Adhering to rocks, and occasionally to floating bodies; has been found attached to Coronula.

Cinaras. Nearly the same habits; there is a group in the Museum of the College of Surgeons adhering to the tail of a water serpent.

Lithotrya. Adhering to the bottom of a deep regular cavity, apparently the work of the animal, in rocks.

Balanus. On rocks and shells at a depth ranging to ten fathoms; affixed to bottoms of ships and other floating bodies.

Octomeris. Attached to rocks.

Conia. Affixed to stones, rocks, shells, &c.

Catophragmus. Attached to Conia, most probably similar habits to those of that genus.

Clitia. Coasts, on shells, &c.

Tubicinella. Imbedded in the blubber of whales.

Coronula. The same.

Chelonobia. On the backs of sea-turtles; adhering to, and sometimes anchored in the tortoise-shell.

Acasta. In sponges.

Creusia. }
Pyrgoma. } Imbedded in corals.

* The observations of Savigny and of Deshayes lead to the conclusion that this genus approaches very closely to the Mollusks, if indeed it does not belong to them. Cuvier, in the last edition of the *Règne Animal*, where he places Dentalium among his Annélides Tubicoles not without hesitation, thus writes: "Si l'opercule rappelle le pied des vermets et des siliquaires, qui déjà ont été transportés dans la classe des mollusques, les branchies rappellent beaucoup celle des amphitrites et des têrebelles. Des observations ultérieures sur leur anatomie, et principalement sur leur système nerveux et vasculaire, resoudront ce problème.—tom. iii. p. 107. W. J. B.

Conchifers.

	Depth in Fathoms.	*Remarks.*
Aspergillum ...	+	In sands. Shallow water probably.
Clavagella	0 to 11	In rocks.
Fistulana.........	+	Sands or hard mud.
Septaria (*Lam.*)	+	Exposed by volcanic action on the coast of Sumatra.
Gastrochæna ...	3 to 10	In the inside of other shells, or in ready-made cavities in rocks, or in cavities of rocks perforated and lined by the animal.
Teredo............	0 to 10	Perforates wood; most destructive to piles, dikes and shipping.
Pholas	0 to 9	Pierces wood, rocks, indurated clay, &c.
Xylophaga	0 to 45	Perforates wood. A specimen thrown up at Gravesend in a piece of stick.
Petricola	0 to 11	In rocks and shells. In cavities of its own working.
Solen	0 to 13	Sandy beaches, wherein it burrows vertically, and lies hidden when the tide is out.
Solenicurtus ...		Moderate depths.
Glycymeris		Probably the same.
Mya..............		Beaches, in which it often lies buried with its tube just projecting. Silt, estuaries.
Panopæa	+	Sands, shallow water.
Anatina		In nearly the same situations.
Lutraria	+	Sands.
Solenella	7 to 45	Soft mud.
Mactra............	0 to 12	Sandy mud and sands.
Galeomma	+	Coasts.
Anatinella	+	Sands. Coast of Ceylon.
Crassatella	8 to 12	Sandy mud.
Pholadomya ...	+	Most probably deep water. The only recent specimen known (which is in my collection,)

	Depth in Fathoms.	*Remarks.*
		was thrown up on the beach at Tortola, after a hurricane.
Solenimya	+	Probably shallow, sands.
Amphidesma ...	0 to 40	Sands and mud.
Cumingia.........	0 to 6	In clay, mud, and in the fissures of rocks.
Corbula	0 to 13	Sandy mud.
Pandora	0 to 10	Sands.
Saxicava		Littoral. In stones and shells.
Venirupis.........		Littoral. In rocks or sands.
Pullastra	0 to 10	Sands, or sandy mud.
Sanguinolaria ...	5 to 13	Sandy mud.
Psammobia	0 to 13	Sands.
Tellina...........	0 to 17	Sands.
Tellinides	5 to 16	Sandy mud.
Corbis	+	Sands probably.
Lucina	5 to 11	Sandy mud and mud.
Ungulina.........	+	Probably sands.
Donax	0 to 10	Sands and sandy mud.
Capsa	5 to 12	Sandy mud and soft mud.
Astarte...........	0 to 10	Sandy mud.
Cyprina	+	Sandy mud.
Cytherea	0 to 50	Mud, sands, coarse sands.
Venus	0 to 50	The same.
Venericardia ...	0 to 50	Mud and sands.
Cardium	0 to 13	Mud, sands and gravel.
Cardita...........	0 to 13	Mud and sands. Sometimes attached to stones.
Cypricardia	+	Sands, and on reefs.
Isocardia	10 to 20	Mud and sand.
Cucullæa	+	Sands.
Byssoarca	0 to 75	Moored to stones and shells.
Arca	0 to 17	Sandy mud and mud, moored to stones, corals, &c.
Pectunculus......	5 to 17	Sandy mud and sands.
Nucula	0 to 60	Sandy mud and sand. Estuary and open sea *.

* According to Mr. Cuming, the same species of this genus vary much in the depths at which they live; for he found *N. cuneata* from 14 to 45 fathoms; *N. obliqua* from 14 to 60 fathoms, and *N. Pisum* from 17 to 45 fathoms. W. J. B.

	Depth in Fathoms.	Remarks.
Trigonia	6 to 14	Only yet discovered near Australia, sandy mud.
Myochama		On Trigoniæ.
Chama............	0 to 17	Affixed to rocks, stones and shells.
Cleidothærus ...		The same. Shallow water.
Tridacna	0 to 7	Moored to rocks, and on coral reefs.
Hippopus.........	0 to 7	Moored to rocks.
Modiola	0 to 17	Littoral. Moored to rocks, stones and shells *.
Mytilus		Littoral. The same, and on Crustaceans, shells, &c.
Lithodomus......	0 to 10	Littoral. Affixed at first by byssus to rocks, which it subsequently penetrates, and remains ever afterwards in the cavity. In shells.
Pinna	0 to 17	Sandy bottoms, moored by byssus.
Crenatula.........	+	In sponges, and moored to corallines, &c.
Perna	0 to 10	Littoral. Moored to mangrove trees, corals, &c.
Malleus	0 to 7	Moored by byssus to rocks, &c.
Avicula	0 to 20	Moored to mangrove trees, corals, shells and rocks.
Meleagrina	0 to 10	Moored by byssus to rocks.
Pedum............	+	Attached by some substance to rocks.
Lima	0 to 30	Moored by byssus.
Pecten............	0 to 20	Sands, sandy mud and mud.
Plicatula	4 to 11	Adheres to stones, shells, &c.
Spondylus	0 to 17	Attached to rocks, corals, &c.
Gryphæa.........	Shallow.	On gravel and sand, estuaries.
Ostrea	0 to 17	On gravel and sand, estuaries and sea. Sometimes attached to rocks, trees, &c. †

* *Modiola discors* floats free, enveloped in its own silky byssus. One Modiola lives in Ascidiæ, and another floats among the Gulf or Sargasso weed. W. J. B.

† There is in my cabinet a good-sized crab, on the back and claws of which are many oysters. Some of the latter are very

	Depth in Fathoms.	*Remarks.*
Vulsella	+	In sponges, &c.
Placuna	+	Sandy bottoms.
Anomia	0 to 12	On oysters and other shells, rocks, &c.
Placunanomia...	11 to 17	Sandy mud.
Crania	+	On stones and shells. Brought up with corals in the Mediterranean, and by cod lines off the coast of Shetland. Probably very deep.
Orbicula	0 to 17	Attached to stones, shells, sunken wrecks, &c.; sandy mud.
Hipponyx	0 to 16	Attached to stones and shells.
Terebratula......	10 to 90	Moored to rocks, shells, &c.
Thecidium		Among red coral. Tuscan seas.
Lingula	0 to 17	Has been found at half tide in hard coarse sand, from four to six inches below the surface of the sand. Coral sands.
	Mollusks.	
Hyalea............	+	Swims freely in the waters.
Chiton...........	0 to 25	Creeps on rocks, stones, &c., to which it adheres.
Capulus (Pileopsis, *Lam.*) ...	0 to 20	Creeps on, and adheres to, shells and stones.
Scutella		Coral sand and sandy beaches.
Patella............	0 to 30	Rocky coasts and stones; seaweeds.
Pleurobranchus.	+	The same.
Umbrella.........	+	The same. Littoral.
Parmophorus ...	+	The same.
Emarginula......	0 to 11	The same.
Siphonaria	+	The same. Littoral.
Fissurella.........	0 to 25	The same.
Calyptræa	0 to 25	Rocks, stones and shells.
Crepidula.........	0 to 40	The same. Sea coasts, estuaries, tidal rivers.
Bulla } Bullæa }	0 to 12	{ Sands and sandy mud. Estuaries.

large, and must have been from six to seven years old. The crab and oysters were alive, when the specimen was brought to London. W. J. B.

	Depth in Fathoms.	*Remarks.*
Aplysia	+	Adheres to rocks.
Dolabella.........	6	Sands.
Melania		Estuaries. Also fresh water.
Nerita		Littoral. Creeps on rocks and sea-weed.
Natica	0 to 40	The same. Mud and sandy mud. Estuaries. Tidal rivers.
Ianthina		Floats freely in the ocean.
Sigaretus	5 to 15	Sand.
Stomatia	7	On meleagrina and corals.
Haliotis		Littoral. Adheres to rocks, &c.
Scissurella	+	
Tornatella	Shallow.	Creeps on sands, leaving furrows.
Pyramidella......	0 to 12	On coral reefs, sands, and sandy mud.
Vermetus.........	0 to 12	In sponges. Under stones, on shells. In coral sand and sand.
Siliquaria.........	+	Has been found in sponges.
Magilus	+	In corals. As the coral increases in volume, the animal secretes a tube, which is almost crystalline and very thick, so as to keep the aperture coequal with the surface of the coral.
Stylifer............	+	Burrows in the rays of Starfish, and found on Echinus, &c. Littoral.
Scalaria	7 to 13	Sandy mud.
Rissoa	+	Coasts and shores.
Delphinula	+	Creeps on rocks and sea-weeds.
Solarium		Littoral. On rocks and weeds.
Rotella............	+	Probably the same.
Trochus	0 to 45	Creeps on rocks and sea-weeds, sand, sandy mud and gravel.
Monodon	+	On rocks and weeds.
Littorina		The same. Littoral *.

* *Littorina pulchra* has been found on mangrove trees 14 feet above the water. They have also been kept alive six months without water.—*Cuming*. W. J. B.

	Depth in Fathoms.	Remarks.
Turbo	0 to 10	On rocks and weeds.
Planaxis		Littoral. Under stones.
Phasianella	Shallow.	Islands, coasts and estuaries.
Turritella.........	5 to 20	Sandy mud.
Cerithium	0 to 17	* Found on various bottoms †. Estuaries.
Potamides		* Estuaries. Also fresh water.
Pleurotoma	8 to 16	* Different bottoms.
Turbinella	0 to 18	* Sandy mud.
Cancellaria	5 to 16	* Sandy mud.
Fasciolaria	0 to 7	* Mud.
Fusus	0 to 11	* Mud, sandy mud and sand.
Pyrula	0 to 9	* The same.
Struthiolaria ...	+	*
Ranella	0 to 11	* Different bottoms.
Murex	5 to 25	* The same.
Typhis	6 to 11	* Sandy mud.
Triton	0 to 30	* Different bottoms.
Rostellaria	+	* A specimen brought up in the mud lying on the fluke of an Indiaman's anchor in the Straits of Macassar.
Pteroceras	+	* As yet only noticed as littoral.
Strombus.........	0 to 13	* Probably different bottoms.
Cassidaria	+	* The same.
Oniscia	+	* Littoral. Coarse sand.
Cassis	5 to 8	* In sands.
Ricinula	+	* On coral reefs and rocks.
Purpura	0 to 25	* The larger proportion of the species of this genus are littoral.

† All those genera distinguished by an asterisk, prefixed to the remarks, may be considered as roaming in pursuit of living prey or dead animal substances; and though they may occur at various depths, and in or on various bottoms, yet they are almost without exception found either on or near the shore or in soundings. The majority penetrate the shells of conchifers by means of an organ which makes a hole as truly round as if it had been cut by an auger, and then suck the juices of the victim.

At least one species of the genus Cerithium is exceedingly tenacious of life; for *C. telescopium*, sent from Calcutta to Mr. G. B. Sowerby in sea water, lived out of water in a small tin box for more than a week.

	Depth in Fathoms.		*Remarks.*
Monoceras	0 to 7	*	On rocks. Nearly all the species littoral.
Concholepas ...		*	Only yet known as littoral.
Harpa	5 to 11	*	Caught by fishing-lines. More frequently by rakes, when out to feed early in the morning.
Dolium	+	*	On reefs.
Nassa	0 to 15	*	Sand, sandy mud, and under stones.
Buccinum	0 to 10	*	Greater part of this genus littoral.
Trichotropis......	10 to 15	*	Bay between Icy Cape and Cape Lisbon.
Eburna	+	*	Sandy mud?
Terebra	0 to 17	*	Sometimes creeps on reefs out of the water, but within reach of the spray.
Columbella	0 to 16	*	Sandy mud and mud.
Mitra	0 to 17	*	Reefs, sandy mud, sands. One species brought up attached to the lead line in the Mediterranean.
Voluta	7 to 14	*	Sands and mud.
Cymba	Shallow.	*	The same.
Melo	Shallow.	*	The same.
Marginella	0 to 9	*	Sand, sandy mud.
Ovulum	0 to 11	*	Under corals and stones, on sea-weeds.
Cypræa		*	Littoral. Under corals and stones.
Terebellum	+	*	
Ancillaria	+	*	One species dredged up in moderately deep water, New Zealand.
Oliva	0 to 12	*	In mud, sandy mud, coarse sand, &c. Caught by fishing-lines.
Conus	0 to 17	*	Sandy mud, &c.
Conohelix	+		On coral reefs.
Conovulus	Shallow.		Marine and estuary.
Nodosaria			Littoral.
Spirula............			Floats on the ocean.

	Depth in Fathoms.	Remarks.
Cristellaria	......	Littoral.
Orbiculina	......	Littoral.
Nautilus	....	Free swimmer and creeps on the bottom.
Argonauta	......	Free swimmer.
Carinaria	....	Near the shore.

THE END.

Printed by Richard Taylor, Red Lion Court, Fleet Street.

Printed in the United States
By Bookmasters